SOLVE
Hidden Numbers

General rules for solving our riddles:

Basic rules
· Each riddle has <u>exactly one solution</u>
· Digits in the solution cannot be repeated
· We always have 5 hints (P1-P5) with strictly defined characteristics

Solution method
Step 1: Analysis of P5 (nothing is correct)
· Hint P5 shows 3 digits that definitely do not appear in the solution
· These digits can be immediately excluded from all positions
Step 2: Analysis of P1 (one digit in correct position)
· P1 contains exactly one digit from the solution in its correct position
· The remaining two digits from P1 can be excluded from the solution

Step 3: Analysis of P2 and P3 (digits in wrong positions)
· P2 and P3 each contain one digit from the solution, but in wrong positions
· The remaining digits from P2 and P3 can be excluded
Step 4: Analysis of P4 (two digits in wrong positions)
· P4 contains two digits from the solution, but in wrong positions
· These digits must come from previous hints (P1, P2, or P3)
· The third digit from P4 does not appear in the solution
Practical Tips
· Systematically cross out digits that definitely do not appear in the solution
· Mark digits that must appear but in different positions
· Check each considered variant against all hints simultaneously

Good luck!

Solution example:

For hints:

P1: 952 (one digit correct and in right position)

P2: 346 (one digit correct but in wrong position)

P3: 071 (one digit correct but in wrong position)

P4: 027 (two digits correct but in wrong positions)

P5: 635 (no digits are correct)

From P5, we know that digits 6,3,5 are not in the solution

From P1, we know that one of the digits [9,5,2] is in the correct position

From P2, we know that one of the digits [3,4,6] is in the solution but in wrong position

etc...

Riddle #

Find answer in #

Solution for #

O.

🔍 23.

→ 908

4	0	2

9̶	5̶	2

One digit is correct and in the right position

6	3	5

No digit is correct

3̶	4	6̶

One digit is correct but in the wrong position

0	2	7̶

Two digits are correct but in wrong positions

0	7̶	1̶

One digit is correct but in the wrong position

1.

🔍 74.
→ 203

2 5 4
One digit is correct and in the right position

2 6 3
No digit is correct

3 8 6
One digit is correct but in the wrong position

7 4 5
Two digits are correct but in wrong positions

1 7 0
One digit is correct but in the wrong position

2.

🔍 189.
→ 986

6 4 0
One digit is correct and in the right position

3 2 6
No digit is correct

9 1 5
One digit is correct but in the wrong position

9 0 8
Two digits are correct but in wrong positions

2 7 3
One digit is correct but in the wrong position

3.

🔍 254.

→ 753

3	2	5

One digit is correct and in the right position

2	9	7

No digit is correct

7	8	9

One digit is correct but in the wrong position

8	4	6

Two digits are correct but in wrong positions

6	0	4

One digit is correct but in the wrong position

4.

🔍 308.

→ 954

0	1	3

One digit is correct and in the right position

0	8	9

No digit is correct

8	2	9

One digit is correct but in the wrong position

2	4	1

Two digits are correct but in wrong positions

5	7	4

One digit is correct but in the wrong position

5.

🔍 373.

→ 162

9	4	2

One digit is correct and in the right position

6	2	0

No digit is correct

0	6	5

One digit is correct but in the wrong position

4	8	7

Two digits are correct but in wrong positions

8	1	7

One digit is correct but in the wrong position

6.

🔍 390.

→ 581

8	3	1

One digit is correct and in the right position

6	2	1

No digit is correct

0	4	5

One digit is correct but in the wrong position

7	4	3

Two digits are correct but in wrong positions

6	2	7

One digit is correct but in the wrong position

7.

🔍 3.

→ 729

9 5 0
One digit is correct and in the right position

9 6 2
No digit is correct

3 1 4
One digit is correct but in the wrong position

5 3 4
Two digits are correct but in wrong positions

2 6 7
One digit is correct but in the wrong position

8.

🔍 314.

→ 215

2 9 1
One digit is correct and in the right position

0 2 4
No digit is correct

7 5 8
One digit is correct but in the wrong position

9 5 6
Two digits are correct but in wrong positions

0 4 3
One digit is correct but in the wrong position

9.

🔍 153.
→ 254

4 1 9
One digit is correct and in the right position

7 3 1
No digit is correct

5 2 0
One digit is correct but in the wrong position

2 8 5
Two digits are correct but in wrong positions

8 3 7
One digit is correct but in the wrong position

10.

🔍 215.
→ 267

8 7 9
One digit is correct and in the right position

5 6 7
No digit is correct

5 2 6
One digit is correct but in the wrong position

9 1 0
Two digits are correct but in wrong positions

3 4 0
One digit is correct but in the wrong position

11.
🔍 290.
→ 638

5	1	6

One digit is correct and in the right position

0	8	1

No digit is correct

3	4	2

One digit is correct but in the wrong position

3	6	4

Two digits are correct but in wrong positions

8	9	0

One digit is correct but in the wrong position

12.
🔍 56.
→ 379

9	7	0

One digit is correct and in the right position

9	2	4

No digit is correct

5	8	1

One digit is correct but in the wrong position

1	0	7

Two digits are correct but in wrong positions

4	6	2

One digit is correct but in the wrong position

13.

🔍 80.
→ 827

9	6	5

One digit is correct and in the right position

7	3	9

No digit is correct

0	8	2

One digit is correct but in the wrong position

5	8	0

Two digits are correct but in wrong positions

3	4	7

One digit is correct but in the wrong position

14.

🔍 368.
→ 823

0	3	5

One digit is correct and in the right position

8	9	3

No digit is correct

1	2	6

One digit is correct but in the wrong position

4	5	1

Two digits are correct but in wrong positions

8	9	7

One digit is correct but in the wrong position

15.

🔍 77.
→ 514

2 1 0
One digit is correct and in the right position

9 5 2
No digit is correct

3 6 4
One digit is correct but in the wrong position

6 0 1
Two digits are correct but in wrong positions

5 9 7
One digit is correct but in the wrong position

16.

🔍 315.
→ 615

5 7 9
One digit is correct and in the right position

2 1 5
No digit is correct

4 2 1
One digit is correct but in the wrong position

9 8 4
Two digits are correct but in wrong positions

0 3 8
One digit is correct but in the wrong position

17.

🔍 5.
→ 137

3	6	8

One digit is
correct and
in the right
position

3	4	7

No digit is correct

2	9	0

One digit is
correct but
in the wrong
position

6	2	9

Two digits are
correct but in
wrong positions

7	4	1

One digit is
correct but
in the wrong
position

18.

🔍 356.
→ 621

0	9	7

One digit is
correct and
in the right
position

1	7	3

No digit is correct

6	8	2

One digit is
correct but
in the wrong
position

5	2	9

Two digits are
correct but in
wrong positions

5	3	1

One digit is
correct but
in the wrong
position

19.
🔍 234.
→ 105

6	1	0

One digit is correct and in the right position

7	6	9

No digit is correct

5	3	8

One digit is correct but in the wrong position

1	2	3

Two digits are correct but in wrong positions

2	9	7

One digit is correct but in the wrong position

20.
🔍 330.
→ 680

4	6	2

One digit is correct and in the right position

7	5	4

No digit is correct

9	8	1

One digit is correct but in the wrong position

2	8	1

Two digits are correct but in wrong positions

5	3	7

One digit is correct but in the wrong position

21.
🔍 65.
→ 693

7 5 9

One digit is correct and in the right position

0 7 4

No digit is correct

4 0 6

One digit is correct but in the wrong position

8 1 5

Two digits are correct but in wrong positions

2 8 1

One digit is correct but in the wrong position

22.
🔍 87.
→ 420

3 0 2

One digit is correct and in the right position

4 1 0

No digit is correct

8 5 6

One digit is correct but in the wrong position

6 2 3

Two digits are correct but in wrong positions

1 4 7

One digit is correct but in the wrong position

23.

🔍 251.

→ 605

8	0	4

One digit is correct and in the right position

4	1	3

No digit is correct

3	6	1

One digit is correct but in the wrong position

7	8	0

Two digits are correct but in wrong positions

5	7	9

One digit is correct but in the wrong position

24.

🔍 381.

→ 183

4	9	1

One digit is correct and in the right position

6	9	0

No digit is correct

3	2	8

One digit is correct but in the wrong position

7	3	5

Two digits are correct but in wrong positions

5	6	0

One digit is correct but in the wrong position

25.

🔍 358.
→ 719

8	4	6

One digit is correct and in the right position

0	4	2

No digit is correct

9	1	3

One digit is correct but in the wrong position

5	7	9

Two digits are correct but in wrong positions

7	2	0

One digit is correct but in the wrong position

26.

🔍 363.
→ 598

2	9	7

One digit is correct and in the right position

1	7	3

No digit is correct

1	8	3

One digit is correct but in the wrong position

9	6	0

Two digits are correct but in wrong positions

6	4	5

One digit is correct but in the wrong position

27.
🔍 274.
→ 970

2	1	0

One digit is correct and in the right position

7	0	8

No digit is correct

4	9	5

One digit is correct but in the wrong position

6	5	1

Two digits are correct but in wrong positions

7	6	8

One digit is correct but in the wrong position

28.
🔍 103.
→ 425

7	8	2

One digit is correct and in the right position

0	7	4

No digit is correct

0	4	1

One digit is correct but in the wrong position

6	9	8

Two digits are correct but in wrong positions

6	3	5

One digit is correct but in the wrong position

29.
🔍 387.
→ 510

5	9	3

One digit is correct and in the right position

6	3	7

No digit is correct

6	7	4

One digit is correct but in the wrong position

8	1	9

Two digits are correct but in wrong positions

8	2	0

One digit is correct but in the wrong position

30.
🔍 14.
→ 718

6	9	3

One digit is correct and in the right position

9	0	5

No digit is correct

4	1	2

One digit is correct but in the wrong position

3	1	2

Two digits are correct but in wrong positions

5	8	0

One digit is correct but in the wrong position

31.

🔍 226.

→ 840

2	1	0

One digit is
correct and
in the right
position

9	6	2

No digit is correct

9	8	6

One digit is
correct but
in the wrong
position

0	7	5

Two digits are
correct but in
wrong positions

4	7	5

One digit is
correct but
in the wrong
position

32.

🔍 331.

→ 491

1	7	2

One digit is
correct and
in the right
position

0	2	8

No digit is correct

9	3	4

One digit is
correct but
in the wrong
position

3	5	7

Two digits are
correct but in
wrong positions

0	6	8

One digit is
correct but
in the wrong
position

33.

🔍 54.

→ 652

9 4 7

One digit is correct and in the right position

2 8 4

No digit is correct

0 2 8

One digit is correct but in the wrong position

7 1 0

Two digits are correct but in wrong positions

6 3 1

One digit is correct but in the wrong position

34.

🔍 229.

→ 759

1 9 8

One digit is correct and in the right position

9 3 6

No digit is correct

6 4 3

One digit is correct but in the wrong position

4 2 0

Two digits are correct but in wrong positions

0 5 7

One digit is correct but in the wrong position

35.

🔍 257.
→ 801

0	5	7

One digit is correct and in the right position

9	0	4

No digit is correct

4	1	9

One digit is correct but in the wrong position

5	2	3

Two digits are correct but in wrong positions

2	8	6

One digit is correct but in the wrong position

36.

🔍 149.
→ 572

5	6	0

One digit is correct and in the right position

6	9	7

No digit is correct

2	3	8

One digit is correct but in the wrong position

1	2	4

Two digits are correct but in wrong positions

7	9	4

One digit is correct but in the wrong position

37.

🔍 242.
→ 749

3	6	0

One digit is correct and in the right position

6	4	2

No digit is correct

2	9	4

One digit is correct but in the wrong position

9	7	1

Two digits are correct but in wrong positions

1	8	7

One digit is correct but in the wrong position

38.

🔍 110.
→ 932

4	5	2

One digit is correct and in the right position

4	9	3

No digit is correct

7	8	0

One digit is correct but in the wrong position

2	7	0

Two digits are correct but in wrong positions

3	1	9

One digit is correct but in the wrong position

39.

🔍 316.
→ 182

4	8	9

One digit is correct and in the right position

2	8	0

No digit is correct

0	3	2

One digit is correct but in the wrong position

3	7	6

Two digits are correct but in wrong positions

6	7	5

One digit is correct but in the wrong position

40.

🔍 92.
→ 347

8	5	2

One digit is correct and in the right position

7	9	8

No digit is correct

3	0	1

One digit is correct but in the wrong position

0	4	5

Two digits are correct but in wrong positions

9	6	7

One digit is correct but in the wrong position

41.
🔍 338.
→ 694

5	6	1

One digit is
correct and
in the right
position

5	3	0

No digit is correct

7	8	9

One digit is
correct but
in the wrong
position

9	1	6

Two digits are
correct but in
wrong positions

0	3	2

One digit is
correct but
in the wrong
position

42.
🔍 75.
→ 740

1	5	0

One digit is
correct and
in the right
position

1	8	4

No digit is correct

9	3	7

One digit is
correct but
in the wrong
position

3	6	5

Two digits are
correct but in
wrong positions

8	4	2

One digit is
correct but
in the wrong
position

43.

🔍 24.

→ 413

4 8 7
One digit is correct and in the right position

5 9 4
No digit is correct

3 0 6
One digit is correct but in the wrong position

3 2 8
Two digits are correct but in wrong positions

5 9 1
One digit is correct but in the wrong position

44.

🔍 397.

→ 518

2 7 4
One digit is correct and in the right position

4 8 3
No digit is correct

3 8 1
One digit is correct but in the wrong position

5 0 7
Two digits are correct but in wrong positions

5 9 6
One digit is correct but in the wrong position

45.
🔍 223.
→ 325

7	5	2

One digit is
correct and
in the right
position

3	5	9

No digit is correct

9	6	3

One digit is
correct but
in the wrong
position

2	1	0

Two digits are
correct but in
wrong positions

1	4	0

One digit is
correct but
in the wrong
position

46.
🔍 361.
→ 935

4	5	6

One digit is
correct and
in the right
position

0	4	9

No digit is correct

0	1	9

One digit is
correct but
in the wrong
position

3	8	5

Two digits are
correct but in
wrong positions

8	2	3

One digit is
correct but
in the wrong
position

47.
🔍 165.
→ 351

0 9 7
One digit is correct and in the right position

1 0 2
No digit is correct

1 2 5
One digit is correct but in the wrong position

4 7 9
Two digits are correct but in wrong positions

3 8 4
One digit is correct but in the wrong position

48.
🔍 128.
→ 583

0 3 7
One digit is correct and in the right position

6 9 0
No digit is correct

9 6 1
One digit is correct but in the wrong position

5 4 3
Two digits are correct but in wrong positions

2 4 8
One digit is correct but in the wrong position

49.

🔍 186.

→ 789

0	7	4

One digit is
correct and
in the right
position

7	1	5

No digit is correct

1	6	5

One digit is
correct but
in the wrong
position

4	9	8

Two digits are
correct but in
wrong positions

8	2	3

One digit is
correct but
in the wrong
position

50.

🔍 313.

→ 419

6	0	8

One digit is
correct and
in the right
position

6	7	2

No digit is correct

1	4	3

One digit is
correct but
in the wrong
position

4	9	0

Two digits are
correct but in
wrong positions

2	7	5

One digit is
correct but
in the wrong
position

51.

🔍 144.
→ 792

8	6	4

One digit is correct and in the right position

5	9	4

No digit is correct

5	9	3

One digit is correct but in the wrong position

3	1	6

Two digits are correct but in wrong positions

1	0	2

One digit is correct but in the wrong position

52.

🔍 253.
→ 842

3	7	1

One digit is correct and in the right position

0	1	2

No digit is correct

4	9	5

One digit is correct but in the wrong position

7	9	4

Two digits are correct but in wrong positions

0	2	8

One digit is correct but in the wrong position

53.

🔍 281.
→ 246

5 3 0
One digit is correct and in the right position

9 5 8
No digit is correct

8 1 9
One digit is correct but in the wrong position

1 6 3
Two digits are correct but in wrong positions

4 6 7
One digit is correct but in the wrong position

54.

🔍 178.
→ 307

5 9 4
One digit is correct and in the right position

9 6 1
No digit is correct

6 7 1
One digit is correct but in the wrong position

7 3 8
Two digits are correct but in wrong positions

0 3 8
One digit is correct but in the wrong position

55.

🔍 256.
→ 962

5 2 0
One digit is correct and in the right position

2 7 8
No digit is correct

4 7 8
One digit is correct but in the wrong position

3 4 6
Two digits are correct but in wrong positions

1 9 6
One digit is correct but in the wrong position

56.

🔍 243.
→ 610

0 5 9
One digit is correct and in the right position

5 8 7
No digit is correct

2 3 4
One digit is correct but in the wrong position

6 9 3
Two digits are correct but in wrong positions

6 8 7
One digit is correct but in the wrong position

57.

🔍 294.
→ 791

3	6	8

One digit is
correct and
in the right
position

1	0	3

No digit is correct

7	5	2

One digit is
correct but
in the wrong
position

2	7	6

Two digits are
correct but in
wrong positions

1	4	0

One digit is
correct but
in the wrong
position

58.

🔍 350.
→ 802

7	4	1

One digit is
correct and
in the right
position

6	1	3

No digit is correct

9	5	2

One digit is
correct but
in the wrong
position

9	0	4

Two digits are
correct but in
wrong positions

3	8	6

One digit is
correct but
in the wrong
position

59.
🔍 309.
→ 317

1 8 9
One digit is correct and in the right position

8 4 7
No digit is correct

2 5 6
One digit is correct but in the wrong position

9 2 6
Two digits are correct but in wrong positions

7 3 4
One digit is correct but in the wrong position

60.
🔍 147.
→ 247

9 4 2
One digit is correct and in the right position

0 9 8
No digit is correct

0 7 8
One digit is correct but in the wrong position

6 5 4
Two digits are correct but in wrong positions

6 1 3
One digit is correct but in the wrong position

61.

🔍 130.

→ 170

8	1	5

One digit is correct and in the right position

9	7	6

One digit is correct but in the wrong position

3	0	5

No digit is correct

6	7	4

Two digits are correct but in wrong positions

4	0	3

One digit is correct but in the wrong position

62.

🔍 8.

→ 392

9	4	5

One digit is correct and in the right position

0	8	1

One digit is correct but in the wrong position

7	9	6

No digit is correct

8	5	1

Two digits are correct but in wrong positions

7	2	6

One digit is correct but in the wrong position

63.
🔍 204.
→ 918

9	3	4

One digit is correct and in the right position

5	0	4

No digit is correct

8	6	1

One digit is correct but in the wrong position

7	2	8

Two digits are correct but in wrong positions

0	2	5

One digit is correct but in the wrong position

64.
🔍 125.
→ 160

0	1	2

One digit is correct and in the right position

0	3	9

No digit is correct

6	4	8

One digit is correct but in the wrong position

8	2	1

Two digits are correct but in wrong positions

3	9	7

One digit is correct but in the wrong position

65.

🔍 117.

→ 658

1 8 7

One digit is correct and in the right position

9 4 7

No digit is correct

9 4 6

One digit is correct but in the wrong position

2 5 8

Two digits are correct but in wrong positions

2 0 5

One digit is correct but in the wrong position

66.

🔍 349.

→ 760

9 0 7

One digit is correct and in the right position

2 3 9

No digit is correct

5 1 8

One digit is correct but in the wrong position

1 7 0

Two digits are correct but in wrong positions

3 2 6

One digit is correct but in the wrong position

67.

Q 137.
→ 635

5	8	6

One digit is
correct and
in the right
position

4	7	8

No digit is correct

4	9	7

One digit is
correct but
in the wrong
position

9	6	0

Two digits are
correct but in
wrong positions

0	3	1

One digit is
correct but
in the wrong
position

68.

Q 353.
→ 138

2	6	1

One digit is
correct and
in the right
position

5	7	2

No digit is correct

9	8	0

One digit is
correct but
in the wrong
position

0	1	6

Two digits are
correct but in
wrong positions

7	5	3

One digit is
correct but
in the wrong
position

69.

🔍 241.
→ 538

5	3	0

One digit is
correct and
in the right
position

3	7	9

No digit is correct

4	1	6

One digit is
correct but
in the wrong
position

2	0	6

Two digits are
correct but in
wrong positions

7	2	9

One digit is
correct but
in the wrong
position

70.

🔍 289.
→ 908

0	2	1

One digit is
correct and
in the right
position

4	6	0

No digit is correct

4	9	6

One digit is
correct but
in the wrong
position

5	3	2

Two digits are
correct but in
wrong positions

5	8	7

One digit is
correct but
in the wrong
position

71.
🔍 337.
→ 154

5	2	1

One digit is correct and in the right position

9	2	0

No digit is correct

9	6	0

One digit is correct but in the wrong position

6	4	3

Two digits are correct but in wrong positions

4	7	3

One digit is correct but in the wrong position

72.
🔍 71.
→ 358

2	7	4

One digit is correct and in the right position

0	7	3

No digit is correct

0	1	3

One digit is correct but in the wrong position

4	9	5

Two digits are correct but in wrong positions

5	8	9

One digit is correct but in the wrong position

73.

🔍 354.
→ 237

7 4 3
One digit is correct and in the right position

0 3 8
No digit is correct

2 6 5
One digit is correct but in the wrong position

4 6 5
Two digits are correct but in wrong positions

0 8 1
One digit is correct but in the wrong position

74.

🔍 94.
→ 857

8 3 6
One digit is correct and in the right position

8 4 7
No digit is correct

7 2 4
One digit is correct but in the wrong position

6 0 3
Two digits are correct but in wrong positions

0 5 9
One digit is correct but in the wrong position

75.

🔍 214.
→ 253

4	8	6

One digit is correct and in the right position

7	9	3

One digit is correct but in the wrong position

7	3	8

No digit is correct

0	6	5

Two digits are correct but in wrong positions

0	1	2

One digit is correct but in the wrong position

- -

76.

🔍 324.
→ 892

7	1	2

One digit is correct and in the right position

9	6	0

One digit is correct but in the wrong position

7	3	4

No digit is correct

0	2	1

Two digits are correct but in wrong positions

4	3	5

One digit is correct but in the wrong position

77.

🔍 348.

→ 716

0	7	8

One digit is
correct and
in the right
position

4	0	3

No digit is correct

4	2	3

One digit is
correct but
in the wrong
position

9	5	7

Two digits are
correct but in
wrong positions

6	9	1

One digit is
correct but
in the wrong
position

78.

🔍 6.

→ 647

3	2	1

One digit is
correct and
in the right
position

3	9	4

No digit is correct

8	7	6

One digit is
correct but
in the wrong
position

7	1	8

Two digits are
correct but in
wrong positions

9	4	5

One digit is
correct but
in the wrong
position

79.
🔍 81.
→ 675

6 5 8
One digit is correct and in the right position

5 0 2
No digit is correct

1 7 4
One digit is correct but in the wrong position

4 3 9
Two digits are correct but in wrong positions

3 0 2
One digit is correct but in the wrong position

80.
🔍 213.
→ 405

8 2 3
One digit is correct and in the right position

4 5 2
No digit is correct

1 9 0
One digit is correct but in the wrong position

7 1 6
Two digits are correct but in wrong positions

4 5 7
One digit is correct but in the wrong position

81.

🔍 325.
→ 643

9 6 1
One digit is correct and in the right position

1 3 7
No digit is correct

4 5 2
One digit is correct but in the wrong position

0 2 6
Two digits are correct but in wrong positions

3 0 7
One digit is correct but in the wrong position

82.

🔍 271.
→ 584

5 9 7
One digit is correct and in the right position

6 7 1
No digit is correct

6 1 4
One digit is correct but in the wrong position

9 0 3
Two digits are correct but in wrong positions

2 0 3
One digit is correct but in the wrong position

83.

🔍 268.
→ 921

4	1	0

One digit is correct and in the right position

7	0	2

No digit is correct

8	5	3

One digit is correct but in the wrong position

1	8	3

Two digits are correct but in wrong positions

7	2	6

One digit is correct but in the wrong position

84.

🔍 224.
→ 713

5	6	8

One digit is correct and in the right position

2	5	3

No digit is correct

9	1	4

One digit is correct but in the wrong position

4	8	7

Two digits are correct but in wrong positions

7	2	3

One digit is correct but in the wrong position

85.

🔍 357.
→ 520

9 7 6
One digit is correct and in the right position

5 9 8
No digit is correct

0 5 8
One digit is correct but in the wrong position

6 1 0
Two digits are correct but in wrong positions

3 2 1
One digit is correct but in the wrong position

86.

🔍 275.
→ 402

5 3 7
One digit is correct and in the right position

8 9 3
No digit is correct

9 1 8
One digit is correct but in the wrong position

6 7 4
Two digits are correct but in wrong positions

6 0 2
One digit is correct but in the wrong position

87.

🔍 108.
→ 762

8	1	6

One digit is
correct and
in the right
position

5	4	1

No digit is correct

5	4	2

One digit is
correct but
in the wrong
position

7	6	8

Two digits are
correct but in
wrong positions

3	0	7

One digit is
correct but
in the wrong
position

88.

🔍 396.
→ 216

0	6	1

One digit is
correct and
in the right
position

0	3	9

No digit is correct

3	8	9

One digit is
correct but
in the wrong
position

7	5	6

Two digits are
correct but in
wrong positions

4	7	2

One digit is
correct but
in the wrong
position

89.

🔍 231.

→ 742

2	4	9

One digit is correct and in the right position

4	0	1

No digit is correct

6	7	5

One digit is correct but in the wrong position

7	9	5

Two digits are correct but in wrong positions

1	8	0

One digit is correct but in the wrong position

90.

🔍 170.

→ 950

9	3	7

One digit is correct and in the right position

2	3	0

No digit is correct

4	2	0

One digit is correct but in the wrong position

4	1	6

Two digits are correct but in wrong positions

6	5	8

One digit is correct but in the wrong position

91.

🔍 27.

→ 412

9	8	1

One digit is correct and in the right position

3	4	8

No digit is correct

6	2	7

One digit is correct but in the wrong position

1	0	7

Two digits are correct but in wrong positions

0	3	4

One digit is correct but in the wrong position

92.

🔍 329.

→ 650

8	6	2

One digit is correct and in the right position

5	4	2

No digit is correct

4	5	7

One digit is correct but in the wrong position

6	9	0

Two digits are correct but in wrong positions

1	9	0

One digit is correct but in the wrong position

93.
🔍 197.
→ 529

2	4	8

One digit is correct and in the right position

5	3	2

No digit is correct

6	3	5

One digit is correct but in the wrong position

8	0	6

Two digits are correct but in wrong positions

9	0	1

One digit is correct but in the wrong position

94.
🔍 83.
→ 230

8	3	1

One digit is correct and in the right position

4	0	3

No digit is correct

0	4	9

One digit is correct but in the wrong position

6	1	2

Two digits are correct but in wrong positions

2	6	5

One digit is correct but in the wrong position

95.

🔍 44.
→ 261

5	4	3

One digit is
correct and
in the right
position

6	9	3

No digit is correct

2	8	0

One digit is
correct but
in the wrong
position

1	8	2

Two digits are
correct but in
wrong positions

6	9	1

One digit is
correct but
in the wrong
position

96.

🔍 230.
→ 375

6	4	7

One digit is
correct and
in the right
position

2	0	7

No digit is correct

8	0	2

One digit is
correct but
in the wrong
position

8	5	9

Two digits are
correct but in
wrong positions

9	1	3

One digit is
correct but
in the wrong
position

97.

🔍 236.
→ 965

4	2	0

One digit is correct and in the right position

7	8	9

One digit is correct but in the wrong position

1	5	2

No digit is correct

3	8	9

Two digits are correct but in wrong positions

5	3	1

One digit is correct but in the wrong position

98.

🔍 359.
→ 709

1	3	5

One digit is correct and in the right position

9	0	4

One digit is correct but in the wrong position

7	6	3

No digit is correct

8	2	9

Two digits are correct but in wrong positions

8	6	7

One digit is correct but in the wrong position

99.
🔍 312.
→ 873

0	7	3

One digit is correct and in the right position

7	1	9

No digit is correct

2	4	5

One digit is correct but in the wrong position

3	8	2

Two digits are correct but in wrong positions

1	9	6

One digit is correct but in the wrong position

100.
🔍 162.
→ 953

2	3	4

One digit is correct and in the right position

5	8	3

No digit is correct

8	0	5

One digit is correct but in the wrong position

0	1	9

Two digits are correct but in wrong positions

1	7	9

One digit is correct but in the wrong position

101.

🔍 155.
→ 603

7	6	2

One digit is correct and in the right position

5	2	9

No digit is correct

8	5	9

One digit is correct but in the wrong position

3	8	0

Two digits are correct but in wrong positions

1	4	3

One digit is correct but in the wrong position

102.

🔍 28.
→ 381

4	0	1

One digit is correct and in the right position

1	3	6

No digit is correct

9	5	7

One digit is correct but in the wrong position

0	5	2

Two digits are correct but in wrong positions

2	3	6

One digit is correct but in the wrong position

103.

🔍 86.
→ 186

6	0	1

One digit is
correct and
in the right
position

3	1	8

No digit is correct

9	4	7

One digit is
correct but
in the wrong
position

2	7	0

Two digits are
correct but in
wrong positions

3	2	8

One digit is
correct but
in the wrong
position

104.

🔍 301.
→ 752

6	2	9

One digit is
correct and
in the right
position

8	3	2

No digit is correct

0	7	5

One digit is
correct but
in the wrong
position

4	9	0

Two digits are
correct but in
wrong positions

3	4	8

One digit is
correct but
in the wrong
position

105.

🔍 366.
→ 703

7 6 9
One digit is correct and in the right position

9 0 5
No digit is correct

5 0 2
One digit is correct but in the wrong position

2 1 4
Two digits are correct but in wrong positions

1 3 4
One digit is correct but in the wrong position

106.

🔍 41.
→ 132

3 8 4
One digit is correct and in the right position

0 3 2
No digit is correct

7 1 9
One digit is correct but in the wrong position

9 4 8
Two digits are correct but in wrong positions

0 2 6
One digit is correct but in the wrong position

107.
🔍 284.
→ 123

1 4 0
One digit is correct and in the right position

6 5 4
No digit is correct

8 9 2
One digit is correct but in the wrong position

8 0 7
Two digits are correct but in wrong positions

6 3 5
One digit is correct but in the wrong position

108.
🔍 364.
→ 276

5 2 4
One digit is correct and in the right position

0 5 9
No digit is correct

7 0 9
One digit is correct but in the wrong position

1 7 2
Two digits are correct but in wrong positions

3 6 1
One digit is correct but in the wrong position

109.
🔍 82.
→ 516

1	8	7

One digit is correct and in the right position

2	9	7

No digit is correct

4	9	2

One digit is correct but in the wrong position

4	6	8

Two digits are correct but in wrong positions

0	5	6

One digit is correct but in the wrong position

110.
🔍 303.
→ 102

2	5	9

One digit is correct and in the right position

3	8	5

No digit is correct

3	4	8

One digit is correct but in the wrong position

4	7	1

Two digits are correct but in wrong positions

6	7	1

One digit is correct but in the wrong position

111.
🔍 295.
→ 695

5	6	1

One digit is correct and in the right position

1	3	8

No digit is correct

8	2	3

One digit is correct but in the wrong position

6	9	5

Two digits are correct but in wrong positions

9	7	0

One digit is correct but in the wrong position

112.
🔍 158.
→ 943

2	1	6

One digit is correct and in the right position

1	3	4

No digit is correct

9	5	8

One digit is correct but in the wrong position

7	6	9

Two digits are correct but in wrong positions

3	7	4

One digit is correct but in the wrong position

113.
🔍 245.
→ 398

9 4 1
One digit is correct and in the right position

4 6 3
No digit is correct

6 5 3
One digit is correct but in the wrong position

5 2 0
Two digits are correct but in wrong positions

7 8 0
One digit is correct but in the wrong position

114.
🔍 278.
→ 501

9 5 1
One digit is correct and in the right position

1 0 4
No digit is correct

7 2 8
One digit is correct but in the wrong position

7 8 5
Two digits are correct but in wrong positions

4 0 3
One digit is correct but in the wrong position

115.

🔍 46.
→ 258

4	3	1

One digit is
correct and
in the right
position

1	7	0

No digit is correct

5	8	6

One digit is
correct but
in the wrong
position

6	5	3

Two digits are
correct but in
wrong positions

7	0	9

One digit is
correct but
in the wrong
position

116.

🔍 111.
→ 651

6	3	7

One digit is
correct and
in the right
position

2	4	7

No digit is correct

5	0	8

One digit is
correct but
in the wrong
position

5	1	9

Two digits are
correct but in
wrong positions

9	4	2

One digit is
correct but
in the wrong
position

117.

🔍 244.
→ 682

7	9	4

One digit is correct and in the right position

0	2	9

No digit is correct

5	6	1

One digit is correct but in the wrong position

4	8	1

Two digits are correct but in wrong positions

0	3	2

One digit is correct but in the wrong position

118.

🔍 51.
→ 284

4	5	2

One digit is correct and in the right position

4	1	8

No digit is correct

9	6	3

One digit is correct but in the wrong position

2	6	9

Two digits are correct but in wrong positions

8	7	1

One digit is correct but in the wrong position

119.
🔍 105.
→ 983

6	0	8

One digit is correct and in the right position

8	5	1

No digit is correct

1	7	5

One digit is correct but in the wrong position

3	4	0

Two digits are correct but in wrong positions

9	3	2

One digit is correct but in the wrong position

120.
🔍 134.
→ 702

6	0	9

One digit is correct and in the right position

8	2	0

No digit is correct

7	5	3

One digit is correct but in the wrong position

4	9	7

Two digits are correct but in wrong positions

2	4	8

One digit is correct but in the wrong position

121.
🔍 120.
→ 940

3	0	1

One digit is
correct and
in the right
position

5	4	1

No digit is correct

4	7	5

One digit is
correct but
in the wrong
position

2	3	0

Two digits are
correct but in
wrong positions

9	2	8

One digit is
correct but
in the wrong
position

122.
🔍 265.
→ 632

5	8	3

One digit is
correct and
in the right
position

8	7	1

No digit is correct

1	6	7

One digit is
correct but
in the wrong
position

6	3	4

Two digits are
correct but in
wrong positions

4	9	2

One digit is
correct but
in the wrong
position

123.

🔍 262.
→ 593

7	1	0

One digit is correct and in the right position

4	3	1

No digit is correct

6	5	8

One digit is correct but in the wrong position

0	2	8

Two digits are correct but in wrong positions

4	9	3

One digit is correct but in the wrong position

124.

🔍 36.
→ 316

6	0	2

One digit is correct and in the right position

8	0	9

No digit is correct

8	5	9

One digit is correct but in the wrong position

2	1	7

Two digits are correct but in wrong positions

3	4	7

One digit is correct but in the wrong position

125.

Q 188.
→ 782

0	6	7

One digit is correct and in the right position

3	8	0

No digit is correct

8	1	3

One digit is correct but in the wrong position

1	9	6

Two digits are correct but in wrong positions

5	9	4

One digit is correct but in the wrong position

126.

Q 270.
→ 981

4	5	0

One digit is correct and in the right position

3	8	0

No digit is correct

6	2	1

One digit is correct but in the wrong position

5	2	1

Two digits are correct but in wrong positions

3	8	9

One digit is correct but in the wrong position

127.

🔍 240.

→ 217

8 6 0
One digit is correct and in the right position

6 9 5
No digit is correct

7 3 1
One digit is correct but in the wrong position

2 0 7
Two digits are correct but in wrong positions

9 4 5
One digit is correct but in the wrong position

128.

🔍 379.

→ 134

1 5 6
One digit is correct and in the right position

3 6 8
No digit is correct

4 0 2
One digit is correct but in the wrong position

9 0 5
Two digits are correct but in wrong positions

8 3 9
One digit is correct but in the wrong position

129.

🔍 198.
→ 795

8	4	2

One digit is
correct and
in the right
position

9	7	2

No digit is correct

1	9	7

One digit is
correct but
in the wrong
position

0	3	1

Two digits are
correct but in
wrong positions

0	5	6

One digit is
correct but
in the wrong
position

130.

🔍 362.
→ 847

9	5	8

One digit is
correct and
in the right
position

9	1	4

No digit is correct

1	3	4

One digit is
correct but
in the wrong
position

2	8	0

Two digits are
correct but in
wrong positions

2	0	6

One digit is
correct but
in the wrong
position

131.

🔍 264.
→ 386

8	1	3

One digit is correct and in the right position

9	1	7

No digit is correct

4	0	6

One digit is correct but in the wrong position

3	5	4

Two digits are correct but in wrong positions

7	9	2

One digit is correct but in the wrong position

132.

🔍 203.
→ 805

7	4	1

One digit is correct and in the right position

2	1	9

No digit is correct

0	5	6

One digit is correct but in the wrong position

0	6	8

Two digits are correct but in wrong positions

8	9	2

One digit is correct but in the wrong position

133.

🔍 391.
→ 527

3 4 1
One digit is correct and in the right position

6 3 0
No digit is correct

6 0 7
One digit is correct but in the wrong position

4 1 5
Two digits are correct but in wrong positions

5 8 9
One digit is correct but in the wrong position

134.

🔍 333.
→ 674

4 3 8
One digit is correct and in the right position

4 9 0
No digit is correct

2 9 0
One digit is correct but in the wrong position

1 8 2
Two digits are correct but in wrong positions

6 1 7
One digit is correct but in the wrong position

135.
Q 305.
→ 142

4	3	1

One digit is correct and in the right position

2	3	7

No digit is correct

5	0	8

One digit is correct but in the wrong position

1	6	5

Two digits are correct but in wrong positions

6	7	2

One digit is correct but in the wrong position

136.
Q 141.
→ 438

5	7	0

One digit is correct and in the right position

7	2	9

No digit is correct

9	1	2

One digit is correct but in the wrong position

1	3	4

Two digits are correct but in wrong positions

3	6	4

One digit is correct but in the wrong position

137.

🔍 304.
→ 509

0	2	1

One digit is
correct and
in the right
position

8	1	9

No digit is correct

7	3	4

One digit is
correct but
in the wrong
position

3	6	2

Two digits are
correct but in
wrong positions

9	5	8

One digit is
correct but
in the wrong
position

138.

🔍 380.
→ 376

6	4	8

One digit is
correct and
in the right
position

4	3	7

No digit is correct

1	2	0

One digit is
correct but
in the wrong
position

9	8	1

Two digits are
correct but in
wrong positions

3	9	7

One digit is
correct but
in the wrong
position

139.
Q 25.
→ 948

7 0 6
One digit is correct and in the right position

6 4 3
No digit is correct

1 4 3
One digit is correct but in the wrong position

9 2 1
Two digits are correct but in wrong positions

8 9 5
One digit is correct but in the wrong position

140.
Q 300.
→ 985

4 3 2
One digit is correct and in the right position

9 7 3
No digit is correct

0 6 8
One digit is correct but in the wrong position

1 0 5
Two digits are correct but in wrong positions

9 7 5
One digit is correct but in the wrong position

141.

🔍 259.
→ 541

1	8	4

One digit is correct and in the right position

5	8	0

No digit is correct

6	7	2

One digit is correct but in the wrong position

4	9	6

Two digits are correct but in wrong positions

5	3	0

One digit is correct but in the wrong position

142.

🔍 319.
→ 239

0	3	4

One digit is correct and in the right position

3	1	2

No digit is correct

2	1	9

One digit is correct but in the wrong position

4	0	7

Two digits are correct but in wrong positions

7	5	8

One digit is correct but in the wrong position

143.

🔍 90.

→ 860

9 6 4
One digit is correct and in the right position

2 4 1
No digit is correct

8 0 7
One digit is correct but in the wrong position

5 0 3
Two digits are correct but in wrong positions

2 1 5
One digit is correct but in the wrong position

144.

🔍 9.

→ 831

7 1 4
One digit is correct and in the right position

8 3 1
No digit is correct

9 0 5
One digit is correct but in the wrong position

4 6 5
Two digits are correct but in wrong positions

3 2 8
One digit is correct but in the wrong position

145.

🔍 129.
→ 273

0 4 5
One digit is correct and in the right position

0 6 8
No digit is correct

9 2 3
One digit is correct but in the wrong position

5 3 9
Two digits are correct but in wrong positions

8 7 6
One digit is correct but in the wrong position

146.

🔍 378.
→ 475

3 7 2
One digit is correct and in the right position

1 4 2
No digit is correct

5 1 4
One digit is correct but in the wrong position

0 6 5
Two digits are correct but in wrong positions

6 9 0
One digit is correct but in the wrong position

147.

🔍 311.
→ 746

1	6	2

One digit is correct and in the right position

5	1	0

No digit is correct

3	4	8

One digit is correct but in the wrong position

3	8	6

Two digits are correct but in wrong positions

0	5	9

One digit is correct but in the wrong position

148.

🔍 62.
→ 685

3	5	7

One digit is correct and in the right position

0	5	4

No digit is correct

6	1	9

One digit is correct but in the wrong position

7	2	9

Two digits are correct but in wrong positions

2	4	0

One digit is correct but in the wrong position

149.

🔍 261.
→ 542

8	5	4

2	3	1

0	7	8

🔍 163.
→ 931

151.
🔍 122.
→ 841

1 3 5

8 1 7
No digit is correct

2 0 4
One digit is correct but in the wrong position

3 2 4
Two digits are correct but in wrong positions

8 7 6
One digit is correct but in the wrong position

152.
🔍 375.
→ 872

7 6 4
One digit is correct and in the right position

0 6 5
No digit is correct

0 8 5
One digit is correct but in the wrong position

4 9 1
Two digits are correct but in wrong positions

3 9 1
One digit is correct but in the wrong position

153.

🔍 78.
→ 458

6	0	2

One digit is correct and in the right position

5	0	8

No digit is correct

4	1	9

One digit is correct but in the wrong position

7	1	4

Two digits are correct but in wrong positions

5	7	8

One digit is correct but in the wrong position

154.

🔍 374.
→ 567

6	4	7

One digit is correct and in the right position

5	6	8

No digit is correct

5	8	3

One digit is correct but in the wrong position

1	9	4

Two digits are correct but in wrong positions

2	1	9

One digit is correct but in the wrong position

155.
🔍 327.
→ 738

0 7 4
One digit is correct and in the right position

1 5 0
No digit is correct

8 9 3
One digit is correct but in the wrong position

2 9 7
Two digits are correct but in wrong positions

1 2 5
One digit is correct but in the wrong position

156.
🔍 367.
→ 508

0 5 1
One digit is correct and in the right position

1 2 8
No digit is correct

6 4 3
One digit is correct but in the wrong position

9 3 5
Two digits are correct but in wrong positions

9 8 2
One digit is correct but in the wrong position

157.

🔍 287.
→ 479

3 6 0
One digit is correct and in the right position

9 4 3
No digit is correct

2 1 7
One digit is correct but in the wrong position

2 0 1
Two digits are correct but in wrong positions

9 8 4
One digit is correct but in the wrong position

158.

🔍 317.
→ 297

0 4 7
One digit is correct and in the right position

2 6 7
No digit is correct

6 2 1
One digit is correct but in the wrong position

4 8 5
Two digits are correct but in wrong positions

8 3 5
One digit is correct but in the wrong position

159.

🔍 219.
→ 135

8	7	0

One digit is
correct and
in the right
position

7	1	3

No digit is correct

2	5	6

One digit is
correct but
in the wrong
position

9	0	6

Two digits are
correct but in
wrong positions

3	1	9

One digit is
correct but
in the wrong
position

160.

🔍 377.
→ 856

6	8	4

One digit is
correct and
in the right
position

9	1	4

No digit is correct

7	9	1

One digit is
correct but
in the wrong
position

0	5	7

Two digits are
correct but in
wrong positions

0	3	2

One digit is
correct but
in the wrong
position

161.
🔍 192.
→ 184

3	2	4

One digit is correct and in the right position

5	6	4

No digit is correct

8	0	1

One digit is correct but in the wrong position

0	7	8

Two digits are correct but in wrong positions

7	6	5

One digit is correct but in the wrong position

162.
🔍 133.
→ 290

8	2	4

One digit is correct and in the right position

4	9	6

No digit is correct

9	6	5

One digit is correct but in the wrong position

2	7	1

Two digits are correct but in wrong positions

0	7	1

One digit is correct but in the wrong position

163.
🔍 97.
→ 245

9	3	0

One digit is
correct and
in the right
position

8	7	0

No digit is correct

8	5	7

One digit is
correct but
in the wrong
position

1	5	6

Two digits are
correct but in
wrong positions

6	2	4

One digit is
correct but
in the wrong
position

164.
🔍 136.
→ 163

5	7	8

One digit is
correct and
in the right
position

1	7	6

No digit is correct

3	9	2

One digit is
correct but
in the wrong
position

9	8	3

Two digits are
correct but in
wrong positions

1	6	4

One digit is
correct but
in the wrong
position

165.

🔍 18.

→ 547

3	5	1

One digit is
correct and
in the right
position

2	9	7

One digit is
correct but
in the wrong
position

3	0	8

No digit is correct

5	1	2

Two digits are
correct but in
wrong positions

8	6	0

One digit is
correct but
in the wrong
position

166.

🔍 168.

→ 204

8	7	6

One digit is
correct and
in the right
position

4	3	9

One digit is
correct but
in the wrong
position

0	8	5

No digit is correct

9	6	2

Two digits are
correct but in
wrong positions

0	5	1

One digit is
correct but
in the wrong
position

167.

🔍 58.
→ 531

5	0	7

One digit is correct and in the right position

7	1	9

No digit is correct

2	6	3

One digit is correct but in the wrong position

0	2	5

Two digits are correct but in wrong positions

9	8	1

One digit is correct but in the wrong position

168.

🔍 185.
→ 196

9	0	2

One digit is correct and in the right position

9	4	3

No digit is correct

8	5	6

One digit is correct but in the wrong position

7	6	0

Two digits are correct but in wrong positions

4	7	3

One digit is correct but in the wrong position

169.

🔍 177.
→ 207

1	6	7

One digit is
correct and
in the right
position

4	7	0

No digit is correct

0	4	5

One digit is
correct but
in the wrong
position

8	2	6

Two digits are
correct but in
wrong positions

8	3	9

One digit is
correct but
in the wrong
position

170.

🔍 237.
→ 964

7	9	1

One digit is
correct and
in the right
position

9	6	5

No digit is correct

0	6	5

One digit is
correct but
in the wrong
position

3	0	8

Two digits are
correct but in
wrong positions

2	8	3

One digit is
correct but
in the wrong
position

171.
🔍 282.
→ 958

4 8 9
One digit is correct and in the right position

6 3 9
No digit is correct

7 1 0
One digit is correct but in the wrong position

5 1 2
Two digits are correct but in wrong positions

6 3 5
One digit is correct but in the wrong position

172.
🔍 383.
→ 418

2 6 3
One digit is correct and in the right position

5 0 6
No digit is correct

5 0 7
One digit is correct but in the wrong position

9 8 7
Two digits are correct but in wrong positions

8 4 1
One digit is correct but in the wrong position

173.
🔍 179.
→ 283

3	5	2

One digit is correct and in the right position

4	2	0

No digit is correct

8	7	6

One digit is correct but in the wrong position

7	3	5

Two digits are correct but in wrong positions

0	1	4

One digit is correct but in the wrong position

174.
🔍 29.
→ 139

5	4	8

One digit is correct and in the right position

3	4	9

No digit is correct

0	3	9

One digit is correct but in the wrong position

2	0	1

Two digits are correct but in wrong positions

1	2	7

One digit is correct but in the wrong position

175.

🔍 266.
→ 704

8	3	0

One digit is correct and in the right position

4	8	2

No digit is correct

4	7	2

One digit is correct but in the wrong position

0	6	1

Two digits are correct but in wrong positions

1	5	6

One digit is correct but in the wrong position

176.

🔍 285.
→ 720

5	3	7

One digit is correct and in the right position

2	7	6

No digit is correct

1	0	9

One digit is correct but in the wrong position

1	8	3

Two digits are correct but in wrong positions

6	4	2

One digit is correct but in the wrong position

177.

🔍 139.
→ 568

3	4	6

One digit is
correct and
in the right
position

6	2	7

No digit is correct

8	1	5

One digit is
correct but
in the wrong
position

4	8	1

Two digits are
correct but in
wrong positions

7	2	9

One digit is
correct but
in the wrong
position

178.

🔍 79.
→ 587

0	7	9

One digit is
correct and
in the right
position

2	8	0

No digit is correct

8	6	2

One digit is
correct but
in the wrong
position

4	5	7

Two digits are
correct but in
wrong positions

5	1	3

One digit is
correct but
in the wrong
position

179.
🔍 98.
→ 157

7	8	6

One digit is correct and in the right position

6	5	3

No digit is correct

4	9	2

One digit is correct but in the wrong position

8	9	0

Two digits are correct but in wrong positions

0	5	3

One digit is correct but in the wrong position

180.
🔍 376.
→ 913

2	5	4

One digit is correct and in the right position

4	7	1

No digit is correct

3	6	8

One digit is correct but in the wrong position

8	0	9

Two digits are correct but in wrong positions

7	0	1

One digit is correct but in the wrong position

181.

🔍 52.

→ 804

6 0 2
One digit is correct and in the right position

6 1 9
No digit is correct

9 1 8
One digit is correct but in the wrong position

4 2 0
Two digits are correct but in wrong positions

7 5 4
One digit is correct but in the wrong position

182.

🔍 206.

→ 689

9 4 1
One digit is correct and in the right position

6 4 8
No digit is correct

8 6 5
One digit is correct but in the wrong position

3 1 2
Two digits are correct but in wrong positions

2 3 0
One digit is correct but in the wrong position

183.

Q 112.
→ 506

6 4 0
One digit is correct and in the right position

1 2 0
No digit is correct

3 2 1
One digit is correct but in the wrong position

7 3 4
Two digits are correct but in wrong positions

5 7 9
One digit is correct but in the wrong position

184.

Q 369.
→ 563

6 1 4
One digit is correct and in the right position

6 7 8
No digit is correct

3 9 5
One digit is correct but in the wrong position

9 5 1
Two digits are correct but in wrong positions

7 8 2
One digit is correct but in the wrong position

185.

🔍 118.
→ 507

9	0	4

One digit is correct and in the right position

8	4	5

Two digits are correct but in wrong positions

7	0	6

No digit is correct

3	2	5

One digit is correct but in the wrong position

6	7	8

One digit is correct but in the wrong position

186.

🔍 22.
→ 684

4	6	5

One digit is correct and in the right position

0	8	2

Two digits are correct but in wrong positions

9	6	3

No digit is correct

8	7	2

One digit is correct but in the wrong position

9	0	3

One digit is correct but in the wrong position

187.

🔍 106.
→ 385

0	3	5

One digit is correct and in the right position

7	0	6

No digit is correct

6	1	7

One digit is correct but in the wrong position

9	2	3

Two digits are correct but in wrong positions

2	4	9

One digit is correct but in the wrong position

188.

🔍 140.
→ 461

7	2	5

One digit is correct and in the right position

4	1	7

No digit is correct

0	3	8

One digit is correct but in the wrong position

5	3	8

Two digits are correct but in wrong positions

4	9	1

One digit is correct but in the wrong position

189.

🔍 370.
→ 790

4	3	5

One digit is correct and in the right position

1	9	3

No digit is correct

9	2	1

One digit is correct but in the wrong position

0	5	7

Two digits are correct but in wrong positions

6	0	7

One digit is correct but in the wrong position

190.

🔍 142.
→ 241

2	8	5

One digit is correct and in the right position

8	4	0

No digit is correct

0	4	3

One digit is correct but in the wrong position

9	6	3

Two digits are correct but in wrong positions

1	9	7

One digit is correct but in the wrong position

191.
🔍 2.
→ 926

9 1 0
One digit is correct and in the right position

7 1 2
No digit is correct

7 6 2
One digit is correct but in the wrong position

6 4 8
Two digits are correct but in wrong positions

8 3 4
One digit is correct but in the wrong position

192.
🔍 174.
→ 387

8 4 9
One digit is correct and in the right position

5 8 7
No digit is correct

3 6 2
One digit is correct but in the wrong position

0 9 3
Two digits are correct but in wrong positions

7 5 1
One digit is correct but in the wrong position

193.

🔍 298.
→ 863

3	6	2

One digit is correct and in the right position

0	8	2

No digit is correct

7	9	1

One digit is correct but in the wrong position

6	9	5

Two digits are correct but in wrong positions

5	8	0

One digit is correct but in the wrong position

194.

🔍 55.
→ 854

9	1	5

One digit is correct and in the right position

1	8	0

No digit is correct

0	2	8

One digit is correct but in the wrong position

2	5	6

Two digits are correct but in wrong positions

4	7	6

One digit is correct but in the wrong position

195.
🔍 93.
→ 701

4 1 9
One digit is correct and in the right position

8 7 4
No digit is correct

2 0 3
One digit is correct but in the wrong position

1 9 2
Two digits are correct but in wrong positions

7 8 5
One digit is correct but in the wrong position

196.
🔍 291.
→ 763

8 3 6
One digit is correct and in the right position

8 0 4
No digit is correct

4 9 0
One digit is correct but in the wrong position

3 6 7
Two digits are correct but in wrong positions

7 2 5
One digit is correct but in the wrong position

197.

🔍 334.
→ 168

1 2 8
One digit is correct and in the right position

1 9 3
No digit is correct

4 5 0
One digit is correct but in the wrong position

8 5 4
Two digits are correct but in wrong positions

3 7 9
One digit is correct but in the wrong position

198.

🔍 190.
→ 810

5 0 1
One digit is correct and in the right position

6 0 7
No digit is correct

4 8 3
One digit is correct but in the wrong position

1 5 4
Two digits are correct but in wrong positions

7 6 2
One digit is correct but in the wrong position

199.

🔍 306.
→ 309

2	1	3

One digit is correct and in the right position

0	6	1

No digit is correct

0	5	6

One digit is correct but in the wrong position

7	3	4

Two digits are correct but in wrong positions

9	7	4

One digit is correct but in the wrong position

200.

🔍 152.
→ 350

6	7	0

One digit is correct and in the right position

4	1	6

No digit is correct

4	1	8

One digit is correct but in the wrong position

2	9	7

Two digits are correct but in wrong positions

2	5	3

One digit is correct but in the wrong position

201.

🔍 57.
→ 370

7	5	0

One digit is correct and in the right position

3	5	2

No digit is correct

3	1	2

One digit is correct but in the wrong position

1	0	9

Two digits are correct but in wrong positions

9	6	4

One digit is correct but in the wrong position

202.

🔍 372.
→ 362

5	2	4

One digit is correct and in the right position

1	9	5

No digit is correct

8	6	7

One digit is correct but in the wrong position

4	6	8

Two digits are correct but in wrong positions

1	3	9

One digit is correct but in the wrong position

203.
🔍 346.
→ 780

2	7	1

One digit is correct and in the right position

2	9	6

No digit is correct

9	6	3

One digit is correct but in the wrong position

7	4	0

Two digits are correct but in wrong positions

4	0	8

One digit is correct but in the wrong position

204.
🔍 388.
→ 982

7	5	2

One digit is correct and in the right position

3	7	1

No digit is correct

1	6	3

One digit is correct but in the wrong position

2	0	9

Two digits are correct but in wrong positions

8	4	9

One digit is correct but in the wrong position

205.

🔍 212.
→ 187

1	6	4

One digit is correct and in the right position

7	9	1

No digit is correct

9	7	2

One digit is correct but in the wrong position

0	4	6

Two digits are correct but in wrong positions

5	0	3

One digit is correct but in the wrong position

206.

🔍 297.
→ 521

1	9	5

One digit is correct and in the right position

5	4	6

No digit is correct

4	0	6

One digit is correct but in the wrong position

0	2	9

Two digits are correct but in wrong positions

7	2	8

One digit is correct but in the wrong position

207.
🔍 194.
→ 426

8	6	0

One digit is correct and in the right position

7	0	9

No digit is correct

4	7	9

One digit is correct but in the wrong position

5	4	3

Two digits are correct but in wrong positions

5	2	1

One digit is correct but in the wrong position

208.
🔍 267.
→ 765

1	4	9

One digit is correct and in the right position

8	9	3

No digit is correct

8	6	3

One digit is correct but in the wrong position

5	1	4

Two digits are correct but in wrong positions

0	5	2

One digit is correct but in the wrong position

209.
🔍 323.
→ 571

2	5	6

One digit is
correct and
in the right
position

2	9	0

No digit is correct

8	1	4

One digit is
correct but
in the wrong
position

8	6	3

Two digits are
correct but in
wrong positions

9	0	7

One digit is
correct but
in the wrong
position

210.
🔍 138.
→ 106

3	9	5

One digit is
correct and
in the right
position

1	8	5

No digit is correct

1	6	8

One digit is
correct but
in the wrong
position

4	6	7

Two digits are
correct but in
wrong positions

7	2	0

One digit is
correct but
in the wrong
position

211.
🔍 202.
→ 639

3	1	9

One digit is
correct and
in the right
position

1	7	5

No digit is correct

6	7	5

One digit is
correct but
in the wrong
position

6	2	0

Two digits are
correct but in
wrong positions

2	8	4

One digit is
correct but
in the wrong
position

212.
🔍 159.
→ 260

1	0	7

One digit is
correct and
in the right
position

0	4	9

No digit is correct

5	9	4

One digit is
correct but
in the wrong
position

5	2	3

Two digits are
correct but in
wrong positions

3	6	8

One digit is
correct but
in the wrong
position

213.

🔍 114.
→ 871

3	6	1

One digit is correct and in the right position

9	8	3

No digit is correct

7	2	0

One digit is correct but in the wrong position

1	4	0

Two digits are correct but in wrong positions

9	5	8

One digit is correct but in the wrong position

214.

🔍 59.
→ 906

3	0	6

One digit is correct and in the right position

9	5	6

No digit is correct

7	4	8

One digit is correct but in the wrong position

7	2	1

Two digits are correct but in wrong positions

1	9	5

One digit is correct but in the wrong position

215.

🔍 73.
→ 209

4	3	6

One digit is
correct and
in the right
position

6	5	8

No digit is correct

1	7	0

One digit is
correct but
in the wrong
position

3	7	9

Two digits are
correct but in
wrong positions

5	8	2

One digit is
correct but
in the wrong
position

216.

🔍 181.
→ 681

1	9	4

One digit is
correct and
in the right
position

7	1	3

No digit is correct

2	5	8

One digit is
correct but
in the wrong
position

5	4	0

Two digits are
correct but in
wrong positions

0	3	7

One digit is
correct but
in the wrong
position

217.

🔍 7.

→ 189

7	0	1

One digit is correct and in the right position

6	1	5

No digit is correct

2	3	4

One digit is correct but in the wrong position

2	9	8

Two digits are correct but in wrong positions

9	5	6

One digit is correct but in the wrong position

218.

🔍 341.

→ 460

5	8	6

One digit is correct and in the right position

1	8	2

No digit is correct

4	2	1

One digit is correct but in the wrong position

3	4	0

Two digits are correct but in wrong positions

7	0	3

One digit is correct but in the wrong position

219.
Q 393.
→ 590

8 1 9
One digit is correct and in the right position

6 7 8
No digit is correct

7 5 6
One digit is correct but in the wrong position

5 4 1
Two digits are correct but in wrong positions

0 4 3
One digit is correct but in the wrong position

220.
Q 67.
→ 120

6 4 2
One digit is correct and in the right position

8 1 4
No digit is correct

3 8 1
One digit is correct but in the wrong position

5 0 3
Two digits are correct but in wrong positions

5 9 7
One digit is correct but in the wrong position

221.
🔍 37.
→ 497

7	3	2

One digit is correct and in the right position

4	5	1

One digit is correct but in the wrong position

2	5	1

No digit is correct

8	9	4

Two digits are correct but in wrong positions

9	6	0

One digit is correct but in the wrong position

222.
🔍 395.
→ 492

1	6	8

One digit is correct and in the right position

3	0	5

One digit is correct but in the wrong position

5	0	8

No digit is correct

9	3	4

Two digits are correct but in wrong positions

9	2	7

One digit is correct but in the wrong position

223.
🔍 23.
→ 602

1 0 2
One digit is
correct and
in the right
position

9 2 3
No digit is correct

3 9 6
One digit is
correct but
in the wrong
position

0 5 8
Two digits are
correct but in
wrong positions

5 8 4
One digit is
correct but
in the wrong
position

224.
🔍 101.
→ 178

6 9 1
One digit is
correct and
in the right
position

5 8 9
No digit is correct

7 3 2
One digit is
correct but
in the wrong
position

3 4 0
Two digits are
correct but in
wrong positions

0 8 5
One digit is
correct but
in the wrong
position

225.

🔍 371.
→ 741

8 7 9
One digit is correct and in the right position

2 5 7
No digit is correct

0 6 1
One digit is correct but in the wrong position

9 6 0
Two digits are correct but in wrong positions

5 4 2
One digit is correct but in the wrong position

226.

🔍 64.
→ 850

4 7 0
One digit is correct and in the right position

2 9 4
No digit is correct

9 2 6
One digit is correct but in the wrong position

6 0 3
Two digits are correct but in wrong positions

8 3 1
One digit is correct but in the wrong position

227.
🔍 384.
→ 973

2	4	1

One digit is
correct and
in the right
position

4	8	5

No digit is correct

7	3	9

One digit is
correct but
in the wrong
position

1	0	7

Two digits are
correct but in
wrong positions

5	6	8

One digit is
correct but
in the wrong
position

228.
🔍 385.
→ 185

5	7	2

One digit is
correct and
in the right
position

2	0	1

No digit is correct

3	4	9

One digit is
correct but
in the wrong
position

8	9	7

Two digits are
correct but in
wrong positions

1	8	0

One digit is
correct but
in the wrong
position

229.
🔍 88.
→ 104

3	1	4

One digit is
correct and
in the right
position

8	4	7

No digit is correct

6	9	0

One digit is
correct but
in the wrong
position

1	6	3

Two digits are
correct but in
wrong positions

7	2	8

One digit is
correct but
in the wrong
position

230.
🔍 246.
→ 698

5	7	3

One digit is
correct and
in the right
position

2	6	7

No digit is correct

2	9	6

One digit is
correct but
in the wrong
position

9	8	1

Two digits are
correct but in
wrong positions

0	4	1

One digit is
correct but
in the wrong
position

231.
Q 104.
→ 859

0	5	9

One digit is
correct and
in the right
position

6	8	9

No digit is correct

6	2	8

One digit is
correct but
in the wrong
position

2	4	5

Two digits are
correct but in
wrong positions

1	7	4

One digit is
correct but
in the wrong
position

232.
Q 135.
→ 586

5	4	6

One digit is
correct and
in the right
position

6	7	8

No digit is correct

2	3	0

One digit is
correct but
in the wrong
position

2	9	4

Two digits are
correct but in
wrong positions

7	1	8

One digit is
correct but
in the wrong
position

233.

🔍 48.
→ 573

7	8	2

One digit is correct and in the right position

6	2	1

No digit is correct

3	0	4

One digit is correct but in the wrong position

8	3	7

Two digits are correct but in wrong positions

1	5	6

One digit is correct but in the wrong position

234.

🔍 199.
→ 812

5	0	6

One digit is correct and in the right position

5	2	7

No digit is correct

4	9	1

One digit is correct but in the wrong position

9	1	0

Two digits are correct but in wrong positions

7	2	3

One digit is correct but in the wrong position

235.
🔍 218.
→ 971

4	9	7

One digit is
correct and
in the right
position

2	5	9

No digit is correct

0	1	8

One digit is
correct but
in the wrong
position

8	0	6

Two digits are
correct but in
wrong positions

6	2	5

One digit is
correct but
in the wrong
position

236.
🔍 398.
→ 493

4	8	7

One digit is
correct and
in the right
position

5	3	4

No digit is correct

5	3	2

One digit is
correct but
in the wrong
position

1	7	8

Two digits are
correct but in
wrong positions

0	1	9

One digit is
correct but
in the wrong
position

237.

🔍 386.
→ 730

5	7	8

One digit is correct and in the right position

7	1	2

No digit is correct

4	1	2

One digit is correct but in the wrong position

3	0	4

Two digits are correct but in wrong positions

0	6	9

One digit is correct but in the wrong position

238.

🔍 296.
→ 687

6	1	8

One digit is correct and in the right position

0	2	1

No digit is correct

3	7	5

One digit is correct but in the wrong position

8	4	3

Two digits are correct but in wrong positions

2	0	9

One digit is correct but in the wrong position

239.
🔍 95.
→ 620

2	8	9

One digit is correct and in the right position

8	4	7

No digit is correct

0	5	6

One digit is correct but in the wrong position

6	1	5

Two digits are correct but in wrong positions

1	4	7

One digit is correct but in the wrong position

240.
🔍 127.
→ 470

3	1	9

One digit is correct and in the right position

9	6	0

No digit is correct

8	7	4

One digit is correct but in the wrong position

4	7	1

Two digits are correct but in wrong positions

6	0	2

One digit is correct but in the wrong position

241.
🔍 11.
→ 562

5	3	4

One digit is
correct and
in the right
position

2	5	1

No digit is correct

9	8	7

One digit is
correct but
in the wrong
position

8	4	3

Two digits are
correct but in
wrong positions

1	6	2

One digit is
correct but
in the wrong
position

242.
🔍 116.
→ 319

6	2	9

One digit is
correct and
in the right
position

0	8	2

No digit is correct

7	4	5

One digit is
correct but
in the wrong
position

1	3	5

Two digits are
correct but in
wrong positions

0	1	8

One digit is
correct but
in the wrong
position

243.

🔍 173.

→ 469

2	9	6

One digit is correct and in the right position

3	4	5

One digit is correct but in the wrong position

4	9	5

No digit is correct

6	3	8

Two digits are correct but in wrong positions

8	1	0

One digit is correct but in the wrong position

244.

🔍 238.

→ 314

4	8	0

One digit is correct and in the right position

0	1	5

No digit is correct

1	5	6

One digit is correct but in the wrong position

7	3	8

Two digits are correct but in wrong positions

9	7	3

One digit is correct but in the wrong position

245.

🔍 217.
→ 905

0	8	7

One digit is correct and in the right position

6	9	8

Two digits are correct but in wrong positions

3	4	0

No digit is correct

4	1	3

One digit is correct but in the wrong position

9	2	5

One digit is correct but in the wrong position

246.

🔍 283.
→ 519

5	9	6

One digit is correct and in the right position

3	2	1

Two digits are correct but in wrong positions

8	6	4

No digit is correct

7	2	1

One digit is correct but in the wrong position

8	4	3

One digit is correct but in the wrong position

247.

🔍 345.

→ 318

6 4 9
One digit is correct and in the right position

2 7 4
Two digits are correct but in wrong positions

8 9 5
No digit is correct

1 2 0
One digit is correct but in the wrong position

8 5 3
One digit is correct but in the wrong position

248.

🔍 123.

→ 942

7 6 3
One digit is correct and in the right position

9 3 4
Two digits are correct but in wrong positions

8 6 2
No digit is correct

8 5 2
One digit is correct but in the wrong position

9 4 1
One digit is correct but in the wrong position

249.

🔍 184.
→ 251

4 1 3
One digit is correct and in the right position

8 1 7
No digit is correct

2 0 6
One digit is correct but in the wrong position

6 3 9
Two digits are correct but in wrong positions

7 5 8
One digit is correct but in the wrong position

250.

🔍 400.
→ 978

3 2 0
One digit is correct and in the right position

8 2 4
No digit is correct

4 7 8
One digit is correct but in the wrong position

7 0 6
Two digits are correct but in wrong positions

5 1 6
One digit is correct but in the wrong position

251.
🔍 151.
→ 607

3 4 6
One digit is correct and in the right position

6 9 5
No digit is correct

2 1 7
One digit is correct but in the wrong position

1 0 4
Two digits are correct but in wrong positions

9 5 8
One digit is correct but in the wrong position

252.
🔍 33.
→ 108

0 7 2
One digit is correct and in the right position

9 0 1
No digit is correct

1 9 6
One digit is correct but in the wrong position

5 2 3
Two digits are correct but in wrong positions

8 4 5
One digit is correct but in the wrong position

253.

🔍 288.
→ 879

4	9	3

One digit is correct and in the right position

1	3	5

No digit is correct

8	6	0

One digit is correct but in the wrong position

6	7	9

Two digits are correct but in wrong positions

5	2	1

One digit is correct but in the wrong position

254.

🔍 32.
→ 368

0	7	1

One digit is correct and in the right position

8	3	7

No digit is correct

3	8	4

One digit is correct but in the wrong position

6	1	9

Two digits are correct but in wrong positions

9	5	2

One digit is correct but in the wrong position

255.

🔍 20.
→ 415

4	8	7

One digit is correct and in the right position

5	7	9

No digit is correct

1	0	3

One digit is correct but in the wrong position

0	3	8

Two digits are correct but in wrong positions

9	5	6

One digit is correct but in the wrong position

256.

🔍 34.
→ 564

0	5	3

One digit is correct and in the right position

6	2	3

No digit is correct

4	9	1

One digit is correct but in the wrong position

9	0	5

Two digits are correct but in wrong positions

6	7	2

One digit is correct but in the wrong position

257.

🔍 50.
→ 152

3	1	6

One digit is correct and in the right position

0	2	3

No digit is correct

7	4	8

One digit is correct but in the wrong position

8	9	1

Two digits are correct but in wrong positions

9	0	2

One digit is correct but in the wrong position

258.

🔍 248.
→ 512

9	3	7

One digit is correct and in the right position

6	3	0

No digit is correct

8	1	4

One digit is correct but in the wrong position

4	2	5

Two digits are correct but in wrong positions

2	6	0

One digit is correct but in the wrong position

259.
🔍 299.
→ 364

4	5	0

One digit is correct and in the right position

8	5	2

No digit is correct

8	3	2

One digit is correct but in the wrong position

3	6	7

Two digits are correct but in wrong positions

9	1	7

One digit is correct but in the wrong position

260.
🔍 121.
→ 286

9	7	6

One digit is correct and in the right position

7	1	8

No digit is correct

1	0	8

One digit is correct but in the wrong position

0	6	4

Two digits are correct but in wrong positions

4	3	5

One digit is correct but in the wrong position

261.
🔍 389.
→ 924

5	0	4

One digit is correct and in the right position

6	2	5

No digit is correct

1	9	3

One digit is correct but in the wrong position

3	9	0

Two digits are correct but in wrong positions

2	6	8

One digit is correct but in the wrong position

262.
🔍 53.
→ 980

8	4	9

One digit is correct and in the right position

0	8	3

No digit is correct

1	7	2

One digit is correct but in the wrong position

6	7	4

Two digits are correct but in wrong positions

0	6	3

One digit is correct but in the wrong position

263.
🔍 102.
→ 430

4	2	1

One digit is correct and in the right position

4	7	0

No digit is correct

7	3	0

One digit is correct but in the wrong position

8	1	5

Two digits are correct but in wrong positions

8	9	6

One digit is correct but in the wrong position

264.
🔍 273.
→ 243

3	7	8

One digit is correct and in the right position

4	5	3

No digit is correct

5	1	4

One digit is correct but in the wrong position

9	8	7

Two digits are correct but in wrong positions

6	9	0

One digit is correct but in the wrong position

265.

🔍 360.

→ 546

8	1	3

One digit is
correct and
in the right
position

3	6	9

No digit is correct

2	9	6

One digit is
correct but
in the wrong
position

2	4	7

Two digits are
correct but in
wrong positions

4	0	5

One digit is
correct but
in the wrong
position

266.

🔍 339.

→ 710

8	4	2

One digit is
correct and
in the right
position

5	4	3

No digit is correct

3	6	5

One digit is
correct but
in the wrong
position

6	9	1

Two digits are
correct but in
wrong positions

1	0	7

One digit is
correct but
in the wrong
position

267.

Q 233.
→ 645

5	2	4

One digit is correct and in the right position

0	6	2

No digit is correct

6	0	7

One digit is correct but in the wrong position

7	3	8

Two digits are correct but in wrong positions

3	1	9

One digit is correct but in the wrong position

268.

Q 171.
→ 618

1	5	0

One digit is correct and in the right position

4	3	0

No digit is correct

7	8	6

One digit is correct but in the wrong position

5	8	7

Two digits are correct but in wrong positions

3	4	9

One digit is correct but in the wrong position

269.

🔍 175.
→ 951

1 2 4
One digit is correct and in the right position

3 1 8
No digit is correct

0 9 5
One digit is correct but in the wrong position

9 4 0
Two digits are correct but in wrong positions

8 3 7
One digit is correct but in the wrong position

270.

🔍 16.
→ 952

8 1 7
One digit is correct and in the right position

0 8 9
No digit is correct

9 0 6
One digit is correct but in the wrong position

7 5 1
Two digits are correct but in wrong positions

5 4 2
One digit is correct but in the wrong position

271.
🔍 355.
→ 490

1	3	2

One digit is
correct and
in the right
position

5	1	7

No digit is correct

5	8	7

One digit is
correct but
in the wrong
position

6	9	3

Two digits are
correct but in
wrong positions

6	0	4

One digit is
correct but
in the wrong
position

272.
🔍 250.
→ 608

4	7	5

One digit is
correct and
in the right
position

4	6	2

No digit is correct

1	8	0

One digit is
correct but
in the wrong
position

8	5	7

Two digits are
correct but in
wrong positions

2	6	9

One digit is
correct but
in the wrong
position

273.
Q 365.
→ 179

3	1	2

One digit is
correct and
in the right
position

8	5	7

One digit is
correct but
in the wrong
position

7	8	1

No digit is correct

2	0	9

Two digits are
correct but in
wrong positions

6	0	9

One digit is
correct but
in the wrong
position

274.
Q 96.
→ 916

3	4	8

One digit is
correct and
in the right
position

0	5	6

One digit is
correct but
in the wrong
position

9	8	2

No digit is correct

6	5	7

Two digits are
correct but in
wrong positions

7	2	9

One digit is
correct but
in the wrong
position

275.
🔍 258.
→ 167

5 7 8
One digit is correct and in the right position

8 0 6
No digit is correct

1 0 6
One digit is correct but in the wrong position

2 9 1
Two digits are correct but in wrong positions

3 2 9
One digit is correct but in the wrong position

276.
🔍 69.
→ 371

0 3 7
One digit is correct and in the right position

0 1 6
No digit is correct

4 5 9
One digit is correct but in the wrong position

3 8 9
Two digits are correct but in wrong positions

8 1 6
One digit is correct but in the wrong position

277.

🔍 249.
→ 726

8	5	4

One digit is
correct and
in the right
position

8	3	6

No digit is correct

1	9	7

One digit is
correct but
in the wrong
position

1	0	5

Two digits are
correct but in
wrong positions

6	2	3

One digit is
correct but
in the wrong
position

278.

🔍 191.
→ 357

9	1	8

One digit is
correct and
in the right
position

5	1	3

No digit is correct

2	0	4

One digit is
correct but
in the wrong
position

2	6	0

Two digits are
correct but in
wrong positions

6	3	5

One digit is
correct but
in the wrong
position

279.
🔍 200.
→ 169

3	6	2

One digit is correct and in the right position

7	6	9

No digit is correct

0	9	7

One digit is correct but in the wrong position

5	0	8

Two digits are correct but in wrong positions

5	8	1

One digit is correct but in the wrong position

280.
🔍 15.
→ 471

6	7	4

One digit is correct and in the right position

6	9	8

No digit is correct

0	2	1

One digit is correct but in the wrong position

1	4	7

Two digits are correct but in wrong positions

9	8	5

One digit is correct but in the wrong position

281.

🔍 160.
→ 731

2 5 9
One digit is correct and in the right position

3 1 9
No digit is correct

6 4 7
One digit is correct but in the wrong position

5 6 0
Two digits are correct but in wrong positions

1 3 8
One digit is correct but in the wrong position

282.

🔍 124.
→ 451

5 1 7
One digit is correct and in the right position

9 2 5
No digit is correct

8 6 4
One digit is correct but in the wrong position

6 0 1
Two digits are correct but in wrong positions

9 2 3
One digit is correct but in the wrong position

283.

🔍 196.

→ 532

4 9 3
One digit is correct and in the right position

6 3 8
Two digits are correct but in wrong positions

5 2 4
No digit is correct

2 7 5
One digit is correct but in the wrong position

1 0 6
One digit is correct but in the wrong position

284.

🔍 169.

→ 380

2 1 6
One digit is correct and in the right position

7 8 0
Two digits are correct but in wrong positions

9 3 1
No digit is correct

3 7 9
One digit is correct but in the wrong position

5 8 0
One digit is correct but in the wrong position

285.

🔍 322.

→ 431

6	2	8

One digit is
correct and
in the right
position

7	0	4

One digit is
correct but
in the wrong
position

1	2	5

No digit is correct

0	3	9

Two digits are
correct but in
wrong positions

9	1	5

One digit is
correct but
in the wrong
position

286.

🔍 45.

→ 298

3	9	6

One digit is
correct and
in the right
position

8	0	2

One digit is
correct but
in the wrong
position

9	4	1

No digit is correct

5	0	2

Two digits are
correct but in
wrong positions

4	5	1

One digit is
correct but
in the wrong
position

287.
🔍 235.
→ 820

4	7	2

One digit is correct and in the right position

1	2	7

Two digits are correct but in wrong positions

5	4	0

No digit is correct

0	5	9

One digit is correct but in the wrong position

6	1	8

One digit is correct but in the wrong position

288.
🔍 336.
→ 296

2	5	9

One digit is correct and in the right position

7	8	5

Two digits are correct but in wrong positions

9	6	1

No digit is correct

3	7	0

One digit is correct but in the wrong position

6	4	1

One digit is correct but in the wrong position

289.
Q 115.
→ 925

2	9	6

One digit is
correct and
in the right
position

0	3	6

No digit is correct

5	0	3

One digit is
correct but
in the wrong
position

8	7	5

Two digits are
correct but in
wrong positions

1	8	4

One digit is
correct but
in the wrong
position

290.
Q 10.
→ 936

8	6	3

One digit is
correct and
in the right
position

3	0	1

No digit is correct

0	1	2

One digit is
correct but
in the wrong
position

6	7	5

Two digits are
correct but in
wrong positions

5	7	9

One digit is
correct but
in the wrong
position

291.

🔍 30.
→ 976

2	4	8

One digit is correct and in the right position

6	4	3

No digit is correct

1	0	5

One digit is correct but in the wrong position

8	5	1

Two digits are correct but in wrong positions

3	7	6

One digit is correct but in the wrong position

292.

🔍 150.
→ 591

0	3	5

One digit is correct and in the right position

7	6	5

No digit is correct

6	7	9

One digit is correct but in the wrong position

3	1	4

Two digits are correct but in wrong positions

1	2	8

One digit is correct but in the wrong position

293.

🔍 225.

→ 706

6 0 1
One digit is correct and in the right position

6 2 9
No digit is correct

2 7 9
One digit is correct but in the wrong position

1 8 4
Two digits are correct but in wrong positions

5 3 4
One digit is correct but in the wrong position

294.

🔍 156.

→ 467

5 1 7
One digit is correct and in the right position

4 1 2
No digit is correct

9 8 3
One digit is correct but in the wrong position

8 6 0
Two digits are correct but in wrong positions

2 4 0
One digit is correct but in the wrong position

295.
🔍 227.
→ 269

5 6 3
One digit is correct and in the right position

5 4 1
No digit is correct

7 2 0
One digit is correct but in the wrong position

2 3 7
Two digits are correct but in wrong positions

1 9 4
One digit is correct but in the wrong position

296.
🔍 89.
→ 938

3 1 2
One digit is correct and in the right position

9 3 5
No digit is correct

4 0 6
One digit is correct but in the wrong position

0 2 4
Two digits are correct but in wrong positions

9 5 7
One digit is correct but in the wrong position

297.
🔍 60.
→ 890

0	4	6

One digit is
correct and
in the right
position

9	5	0

No digit is correct

7	8	1

One digit is
correct but
in the wrong
position

4	7	3

Two digits are
correct but in
wrong positions

5	9	2

One digit is
correct but
in the wrong
position

298.
🔍 347.
→ 359

5	4	0

One digit is
correct and
in the right
position

0	1	9

No digit is correct

6	8	2

One digit is
correct but
in the wrong
position

7	8	4

Two digits are
correct but in
wrong positions

1	9	7

One digit is
correct but
in the wrong
position

299.

🔍 31.
→ 473

9 4 5
One digit is correct and in the right position

9 7 3
No digit is correct

6 0 1
One digit is correct but in the wrong position

1 0 4
Two digits are correct but in wrong positions

7 3 8
One digit is correct but in the wrong position

300.

🔍 326.
→ 450

1 4 3
One digit is correct and in the right position

6 5 4
No digit is correct

0 2 7
One digit is correct but in the wrong position

9 8 0
Two digits are correct but in wrong positions

6 9 5
One digit is correct but in the wrong position

301.

🔍 4.
→ 604

3	8	4

One digit is correct and in the right position

8	1	6

No digit is correct

6	9	1

One digit is correct but in the wrong position

4	0	5

Two digits are correct but in wrong positions

7	2	5

One digit is correct but in the wrong position

302.

🔍 216.
→ 570

6	2	5

One digit is correct and in the right position

0	9	2

No digit is correct

8	7	3

One digit is correct but in the wrong position

1	5	8

Two digits are correct but in wrong positions

9	1	0

One digit is correct but in the wrong position

303.
Q 201.
→ 214

3	6	1

One digit is correct and in the right position

6	8	2

No digit is correct

0	8	2

One digit is correct but in the wrong position

0	9	7

Two digits are correct but in wrong positions

7	5	4

One digit is correct but in the wrong position

304.
Q 260.
→ 523

2	3	9

One digit is correct and in the right position

9	1	0

No digit is correct

4	6	7

One digit is correct but in the wrong position

6	7	8

Two digits are correct but in wrong positions

8	1	0

One digit is correct but in the wrong position

305.

Q 340.
→ 456

9	8	4

One digit is correct and in the right position

9	7	2

No digit is correct

0	5	6

One digit is correct but in the wrong position

3	4	6

Two digits are correct but in wrong positions

7	1	2

One digit is correct but in the wrong position

306.

Q 195.
→ 543

8	5	1

One digit is correct and in the right position

2	3	5

No digit is correct

0	9	6

One digit is correct but in the wrong position

1	8	0

Two digits are correct but in wrong positions

2	3	7

One digit is correct but in the wrong position

307.
🔍 321.
→ 817

4 2 3
One digit is correct and in the right position

4 7 0
No digit is correct

7 0 1
One digit is correct but in the wrong position

5 8 2
Two digits are correct but in wrong positions

5 6 9
One digit is correct but in the wrong position

308.
🔍 143.
→ 712

8 2 4
One digit is correct and in the right position

7 2 3
No digit is correct

6 7 3
One digit is correct but in the wrong position

5 0 6
Two digits are correct but in wrong positions

0 1 5
One digit is correct but in the wrong position

309.
Q 205.
→ 369

3	9	7

One digit is
correct and
in the right
position

4	3	5

No digit is correct

5	4	1

One digit is
correct but
in the wrong
position

8	7	9

Two digits are
correct but in
wrong positions

0	2	8

One digit is
correct but
in the wrong
position

310.
Q 43.
→ 807

7	5	3

One digit is
correct and
in the right
position

5	2	0

No digit is correct

1	6	8

One digit is
correct but
in the wrong
position

6	3	1

Two digits are
correct but in
wrong positions

0	2	4

One digit is
correct but
in the wrong
position

311.
🔍 352.
→ 963

6 5 0
One digit is correct and in the right position

4 2 5
No digit is correct

7 2 4
One digit is correct but in the wrong position

7 9 1
Two digits are correct but in wrong positions

9 3 8
One digit is correct but in the wrong position

312.
🔍 47.
→ 623

4 0 1
One digit is correct and in the right position

7 9 4
No digit is correct

7 9 3
One digit is correct but in the wrong position

5 1 2
Two digits are correct but in wrong positions

5 8 6
One digit is correct but in the wrong position

313.
🔍 209.
→ 504

5 4 9
One digit is
correct and
in the right
position

8 6 4
No digit is correct

6 1 8
One digit is
correct but
in the wrong
position

1 9 7
Two digits are
correct but in
wrong positions

7 3 2
One digit is
correct but
in the wrong
position

314.
🔍 39.
→ 395

1 5 9
One digit is
correct and
in the right
position

6 5 0
No digit is correct

6 2 0
One digit is
correct but
in the wrong
position

2 9 8
Two digits are
correct but in
wrong positions

7 4 8
One digit is
correct but
in the wrong
position

315.
🔍 286.
→ 349

1	5	8

One digit is correct and in the right position

3	7	1

No digit is correct

9	0	4

One digit is correct but in the wrong position

8	4	9

Two digits are correct but in wrong positions

3	2	7

One digit is correct but in the wrong position

316.
🔍 392.
→ 463

1	7	8

One digit is correct and in the right position

3	0	1

No digit is correct

6	9	5

One digit is correct but in the wrong position

9	8	6

Two digits are correct but in wrong positions

0	4	3

One digit is correct but in the wrong position

317.

Q 208.
→ 148

8	9	5

One digit is correct and in the right position

4	5	6

Two digits are correct but in wrong positions

1	0	8

No digit is correct

6	4	3

One digit is correct but in the wrong position

1	0	7

One digit is correct but in the wrong position

318.

Q 166.
→ 574

2	6	7

One digit is correct and in the right position

4	1	0

Two digits are correct but in wrong positions

6	8	9

No digit is correct

8	9	0

One digit is correct but in the wrong position

5	4	3

One digit is correct but in the wrong position

319.
Q 210.
→ 974

1 4 8
One digit is correct and in the right position

4 7 3
No digit is correct

5 6 9
One digit is correct but in the wrong position

6 2 0
Two digits are correct but in wrong positions

3 7 0
One digit is correct but in the wrong position

320.
Q 19.
→ 907

8 0 3
One digit is correct and in the right position

3 4 9
No digit is correct

2 7 1
One digit is correct but in the wrong position

0 5 7
Two digits are correct but in wrong positions

5 9 4
One digit is correct but in the wrong position

321.
🔍 232.
→ 125

0	9	6

One digit is correct and in the right position

7	3	0

No digit is correct

8	1	4

One digit is correct but in the wrong position

2	6	8

Two digits are correct but in wrong positions

7	3	5

One digit is correct but in the wrong position

322.
🔍 239.
→ 690

6	5	9

One digit is correct and in the right position

9	1	7

No digit is correct

8	0	3

One digit is correct but in the wrong position

0	3	2

Two digits are correct but in wrong positions

2	7	1

One digit is correct but in the wrong position

323.

🔍 183.
→ 786

9 0 4
One digit is correct and in the right position

2 8 9
No digit is correct

8 2 5
One digit is correct but in the wrong position

6 1 0
Two digits are correct but in wrong positions

6 3 1
One digit is correct but in the wrong position

324.

🔍 145.
→ 502

6 7 0
One digit is correct and in the right position

0 5 1
No digit is correct

5 1 2
One digit is correct but in the wrong position

3 9 7
Two digits are correct but in wrong positions

4 3 9
One digit is correct but in the wrong position

325.

Q 277.
→ 560

9 2 0
One digit is correct and in the right position

9 8 4
No digit is correct

5 6 3
One digit is correct but in the wrong position

6 0 2
Two digits are correct but in wrong positions

8 7 4
One digit is correct but in the wrong position

326.

Q 61.
→ 109

8 3 0
One digit is correct and in the right position

9 3 4
No digit is correct

9 1 4
One digit is correct but in the wrong position

0 2 7
Two digits are correct but in wrong positions

2 5 7
One digit is correct but in the wrong position

327.
🔍 172.
→ 372

4	7	9

One digit is correct and in the right position

3	6	7

No digit is correct

1	2	0

One digit is correct but in the wrong position

9	8	1

Two digits are correct but in wrong positions

8	3	6

One digit is correct but in the wrong position

328.
🔍 222.
→ 725

8	9	7

One digit is correct and in the right position

3	8	0

No digit is correct

2	1	5

One digit is correct but in the wrong position

2	5	9

Two digits are correct but in wrong positions

0	3	4

One digit is correct but in the wrong position

329.

🔍 68.
→ 769

1	6	9

One digit is correct and in the right position

7	6	0

No digit is correct

5	4	3

One digit is correct but in the wrong position

2	8	3

Two digits are correct but in wrong positions

8	0	7

One digit is correct but in the wrong position

330.

🔍 247.
→ 312

3	5	0

One digit is correct and in the right position

5	6	9

No digit is correct

1	2	7

One digit is correct but in the wrong position

1	8	4

Two digits are correct but in wrong positions

8	6	9

One digit is correct but in the wrong position

331.
🔍 272.
→ 673

4	0	2

One digit is
correct and
in the right
position

9	7	4

No digit is correct

8	1	5

One digit is
correct but
in the wrong
position

2	8	0

Two digits are
correct but in
wrong positions

9	7	6

One digit is
correct but
in the wrong
position

332.
🔍 276.
→ 103

8	9	1

One digit is
correct and
in the right
position

0	8	2

No digit is correct

7	2	0

One digit is
correct but
in the wrong
position

1	6	7

Two digits are
correct but in
wrong positions

4	3	6

One digit is
correct but
in the wrong
position

333.
Q 344.
→ 728

1 0 2
One digit is correct and in the right position

3 5 1
No digit is correct

9 6 7
One digit is correct but in the wrong position

6 2 7
Two digits are correct but in wrong positions

5 4 3
One digit is correct but in the wrong position

334.
Q 180.
→ 748

9 4 0
One digit is correct and in the right position

5 0 2
No digit is correct

1 5 2
One digit is correct but in the wrong position

6 3 1
Two digits are correct but in wrong positions

7 3 6
One digit is correct but in the wrong position

335.

🔍 318.

→ 197

1	7	9

One digit is correct and in the right position

1	0	2

No digit is correct

4	6	8

One digit is correct but in the wrong position

4	8	7

Two digits are correct but in wrong positions

2	0	5

One digit is correct but in the wrong position

336.

🔍 40.

→ 457

2	1	7

One digit is correct and in the right position

6	1	9

No digit is correct

6	3	9

One digit is correct but in the wrong position

7	8	4

Two digits are correct but in wrong positions

4	0	5

One digit is correct but in the wrong position

337.

Q 307.
→ 536

3	1	6

One digit is correct and in the right position

2	9	6

No digit is correct

7	5	4

One digit is correct but in the wrong position

7	4	1

Two digits are correct but in wrong positions

2	9	8

One digit is correct but in the wrong position

338.

Q 280.
→ 291

2	9	1

One digit is correct and in the right position

5	8	9

No digit is correct

5	8	4

One digit is correct but in the wrong position

1	2	7

Two digits are correct but in wrong positions

7	6	0

One digit is correct but in the wrong position

339.

🔍 49.

→ 816

6	8	0

One digit is correct and in the right position

2	5	7

One digit is correct but in the wrong position

0	5	2

No digit is correct

9	4	8

Two digits are correct but in wrong positions

9	1	3

One digit is correct but in the wrong position

340.

🔍 164.

→ 164

1	2	9

One digit is correct and in the right position

0	3	4

One digit is correct but in the wrong position

8	7	9

No digit is correct

5	3	6

Two digits are correct but in wrong positions

6	7	8

One digit is correct but in the wrong position

341.

🔍 252.
→ 534

9 0 4
One digit is correct and in the right position

7 2 9
No digit is correct

7 1 2
One digit is correct but in the wrong position

8 4 0
Two digits are correct but in wrong positions

6 8 3
One digit is correct but in the wrong position

342.

🔍 320.
→ 218

6 5 7
One digit is correct and in the right position

1 6 3
No digit is correct

0 8 2
One digit is correct but in the wrong position

7 8 0
Two digits are correct but in wrong positions

3 9 1
One digit is correct but in the wrong position

343.
🔍 119.
→ 513

5	8	6

One digit is
correct and
in the right
position

4	7	6

No digit is correct

4	7	9

One digit is
correct but
in the wrong
position

3	1	8

Two digits are
correct but in
wrong positions

0	3	2

One digit is
correct but
in the wrong
position

344.
🔍 302.
→ 472

4	1	0

One digit is
correct and
in the right
position

3	2	1

No digit is correct

2	5	3

One digit is
correct but
in the wrong
position

8	0	7

Two digits are
correct but in
wrong positions

7	8	9

One digit is
correct but
in the wrong
position

345.
🔍 176.
→ 342

8	9	0

One digit is correct and in the right position

9	1	5

No digit is correct

5	1	7

One digit is correct but in the wrong position

0	6	2

Two digits are correct but in wrong positions

2	4	3

One digit is correct but in the wrong position

346.
🔍 193.
→ 374

0	6	9

One digit is correct and in the right position

2	9	7

No digit is correct

2	3	7

One digit is correct but in the wrong position

3	5	6

Two digits are correct but in wrong positions

4	8	5

One digit is correct but in the wrong position

347.

🔍 255.
→ 578

7 0 5
One digit is correct and in the right position

8 4 9
One digit is correct but in the wrong position

9 8 0
No digit is correct

6 5 1
Two digits are correct but in wrong positions

1 2 3
One digit is correct but in the wrong position

348.

🔍 148.
→ 279

6 4 1
One digit is correct and in the right position

7 0 8
One digit is correct but in the wrong position

7 0 1
No digit is correct

8 5 2
Two digits are correct but in wrong positions

5 9 3
One digit is correct but in the wrong position

349.

🔍 72.

→ 601

3	4	6

One digit is
correct and
in the right
position

7	6	0

No digit is correct

2	8	9

One digit is
correct but
in the wrong
position

8	1	5

Two digits are
correct but in
wrong positions

5	7	0

One digit is
correct but
in the wrong
position

350.

🔍 221.

→ 849

4	2	3

One digit is
correct and
in the right
position

5	1	2

No digit is correct

9	6	8

One digit is
correct but
in the wrong
position

7	0	9

Two digits are
correct but in
wrong positions

1	7	5

One digit is
correct but
in the wrong
position

351.
🔍 21.
→ 756

5 0 3
One digit is correct and in the right position

8 7 5
No digit is correct

4 1 9
One digit is correct but in the wrong position

3 4 9
Two digits are correct but in wrong positions

8 6 7
One digit is correct but in the wrong position

352.
🔍 13.
→ 679

5 4 7
One digit is correct and in the right position

0 3 4
No digit is correct

3 0 8
One digit is correct but in the wrong position

7 5 2
Two digits are correct but in wrong positions

2 1 9
One digit is correct but in the wrong position

353.

🔍 228.
→ 301

1 6 0
One digit is correct and in the right position

7 9 6
No digit is correct

5 7 9
One digit is correct but in the wrong position

2 5 8
Two digits are correct but in wrong positions

8 2 3
One digit is correct but in the wrong position

354.

🔍 207.
→ 146

5 3 6
One digit is correct and in the right position

1 3 8
No digit is correct

1 4 8
One digit is correct but in the wrong position

6 0 2
Two digits are correct but in wrong positions

2 0 7
One digit is correct but in the wrong position

355.

🔍 107.

→ 836

1	6	0

One digit is correct and in the right position

4	5	6

No digit is correct

4	3	5

One digit is correct but in the wrong position

3	9	2

Two digits are correct but in wrong positions

8	7	2

One digit is correct but in the wrong position

356.

🔍 42.

→ 895

7	9	3

One digit is correct and in the right position

8	5	3

No digit is correct

4	6	2

One digit is correct but in the wrong position

0	1	4

Two digits are correct but in wrong positions

0	8	5

One digit is correct but in the wrong position

357.

🔍 332.
→ 206

1	2	8

One digit is correct and in the right position

5	7	8

No digit is correct

7	5	0

One digit is correct but in the wrong position

0	3	9

Two digits are correct but in wrong positions

6	3	9

One digit is correct but in the wrong position

358.

🔍 109.
→ 897

7	1	8

One digit is correct and in the right position

3	7	4

No digit is correct

3	4	5

One digit is correct but in the wrong position

6	2	1

Two digits are correct but in wrong positions

6	9	2

One digit is correct but in the wrong position

359.
🔍 182.
→ 198

5 0 9
One digit is correct and in the right position

4 1 5
No digit is correct

2 7 6
One digit is correct but in the wrong position

8 9 7
Two digits are correct but in wrong positions

4 1 8
One digit is correct but in the wrong position

360.
🔍 126.
→ 824

2 3 1
One digit is correct and in the right position

6 7 3
No digit is correct

7 9 6
One digit is correct but in the wrong position

5 1 8
Two digits are correct but in wrong positions

0 5 8
One digit is correct but in the wrong position

361.

🔍 211.
→ 158

6 1 0
One digit is correct and in the right position

3 9 2
Two digits are correct but in wrong positions

4 0 7
No digit is correct

5 8 3
One digit is correct but in the wrong position

7 9 4
One digit is correct but in the wrong position

362.

🔍 70.
→ 328

1 0 7
One digit is correct and in the right position

8 7 0
Two digits are correct but in wrong positions

2 1 6
No digit is correct

4 8 5
One digit is correct but in the wrong position

6 2 9
One digit is correct but in the wrong position

363.
🔍 85.
→ 896

4 2 8
One digit is correct and in the right position

4 9 7
No digit is correct

7 9 5
One digit is correct but in the wrong position

0 3 2
Two digits are correct but in wrong positions

6 0 1
One digit is correct but in the wrong position

364.
🔍 84.
→ 627

0 1 4
One digit is correct and in the right position

4 9 8
No digit is correct

9 8 7
One digit is correct but in the wrong position

5 3 1
Two digits are correct but in wrong positions

6 3 2
One digit is correct but in the wrong position

365.

🔍 394.
→ 592

6 8 7
One digit is correct and in the right position

3 8 0
No digit is correct

9 4 5
One digit is correct but in the wrong position

2 9 1
Two digits are correct but in wrong positions

0 3 2
One digit is correct but in the wrong position

366.

🔍 310.
→ 721

1 6 7
One digit is correct and in the right position

3 2 1
No digit is correct

5 9 0
One digit is correct but in the wrong position

0 7 6
Two digits are correct but in wrong positions

2 8 3
One digit is correct but in the wrong position

367.
🔍 263.
→ 459

1 8 0
One digit is correct and in the right position

2 8 5
No digit is correct

3 5 2
One digit is correct but in the wrong position

0 7 3
Two digits are correct but in wrong positions

9 4 7
One digit is correct but in the wrong position

368.
🔍 167.
→ 715

5 9 8
One digit is correct and in the right position

0 6 8
No digit is correct

3 0 6
One digit is correct but in the wrong position

2 1 3
Two digits are correct but in wrong positions

7 1 4
One digit is correct but in the wrong position

369.
Q 91.
→ 219

4 0 6
One digit is correct and in the right position

3 8 0
No digit is correct

9 7 1
One digit is correct but in the wrong position

1 2 7
Two digits are correct but in wrong positions

2 8 3
One digit is correct but in the wrong position

370.
Q 161.
→ 275

9 8 2
One digit is correct and in the right position

9 5 3
No digit is correct

7 4 6
One digit is correct but in the wrong position

4 2 8
Two digits are correct but in wrong positions

3 1 5
One digit is correct but in the wrong position

371.
🔍 66.
→ 409

7 3 1
One digit is correct and in the right position

9 2 6
One digit is correct but in the wrong position

4 5 3
No digit is correct

0 9 6
Two digits are correct but in wrong positions

4 0 5
One digit is correct but in the wrong position

372.
🔍 279.
→ 384

0 6 7
One digit is correct and in the right position

4 1 8
One digit is correct but in the wrong position

4 8 7
No digit is correct

6 9 2
Two digits are correct but in wrong positions

9 3 5
One digit is correct but in the wrong position

373.

Q 26.
→ 548

5	7	4

One digit is correct and in the right position

8	2	1

One digit is correct but in the wrong position

4	2	1

No digit is correct

8	3	9

Two digits are correct but in wrong positions

9	6	0

One digit is correct but in the wrong position

374.

Q 187.
→ 341

6	4	5

One digit is correct and in the right position

8	9	7

One digit is correct but in the wrong position

2	6	0

No digit is correct

5	1	8

Two digits are correct but in wrong positions

2	3	0

One digit is correct but in the wrong position

375.
Q 35.
→ 814

9 0 6
One digit is correct and in the right position

3 9 7
No digit is correct

4 1 5
One digit is correct but in the wrong position

1 6 0
Two digits are correct but in wrong positions

3 8 7
One digit is correct but in the wrong position

376.
Q 154.
→ 280

9 0 7
One digit is correct and in the right position

8 9 4
No digit is correct

8 5 4
One digit is correct but in the wrong position

7 2 6
Two digits are correct but in wrong positions

3 1 6
One digit is correct but in the wrong position

377.
🔍 100.
→ 670

9	6	1

One digit is correct and in the right position

1	2	0

No digit is correct

5	0	2

One digit is correct but in the wrong position

3	8	5

Two digits are correct but in wrong positions

7	3	4

One digit is correct but in the wrong position

378.
🔍 342.
→ 356

0	7	8

One digit is correct and in the right position

9	7	3

No digit is correct

1	6	4

One digit is correct but in the wrong position

8	4	1

Two digits are correct but in wrong positions

9	2	3

One digit is correct but in the wrong position

379.
🔍 328.
→ 190

3	2	6

One digit is
correct and
in the right
position

1	3	9

No digit is correct

5	8	4

One digit is
correct but
in the wrong
position

5	0	2

Two digits are
correct but in
wrong positions

9	1	7

One digit is
correct but
in the wrong
position

380.
🔍 113.
→ 619

1	9	4

One digit is
correct and
in the right
position

2	6	1

No digit is correct

8	5	0

One digit is
correct but
in the wrong
position

8	7	9

Two digits are
correct but in
wrong positions

2	6	3

One digit is
correct but
in the wrong
position

381.

Q 220.
→ 453

9 2 7
One digit is correct and in the right position

6 4 9
No digit is correct

6 4 1
One digit is correct but in the wrong position

0 3 2
Two digits are correct but in wrong positions

5 0 8
One digit is correct but in the wrong position

382.

Q 351.
→ 406

4 5 1
One digit is correct and in the right position

9 3 4
No digit is correct

9 3 7
One digit is correct but in the wrong position

6 0 5
Two digits are correct but in wrong positions

8 6 2
One digit is correct but in the wrong position

383.
🔍 335.
→ 278

6 2 7
One digit is correct and in the right position

5 6 3
No digit is correct

4 8 9
One digit is correct but in the wrong position

9 7 0
Two digits are correct but in wrong positions

5 3 1
One digit is correct but in the wrong position

384.
🔍 146.
→ 671

9 6 5
One digit is correct and in the right position

6 8 1
No digit is correct

8 4 1
One digit is correct but in the wrong position

0 5 7
Two digits are correct but in wrong positions

3 2 7
One digit is correct but in the wrong position

385.
🔍 131.
→ 478

3 5 7
One digit is correct and in the right position

7 1 4
No digit is correct

8 1 4
One digit is correct but in the wrong position

5 6 8
Two digits are correct but in wrong positions

6 2 0
One digit is correct but in the wrong position

386.
🔍 399.
→ 540

9 6 0
One digit is correct and in the right position

5 6 7
No digit is correct

5 4 7
One digit is correct but in the wrong position

0 3 8
Two digits are correct but in wrong positions

8 2 1
One digit is correct but in the wrong position

387.
🔍 12.
→ 498

1	5	9

One digit is
correct and
in the right
position

4	1	0

No digit is correct

7	8	2

One digit is
correct but
in the wrong
position

7	9	6

Two digits are
correct but in
wrong positions

4	3	0

One digit is
correct but
in the wrong
position

388.
🔍 1.
→ 692

6	7	3

One digit is
correct and
in the right
position

4	6	8

No digit is correct

0	9	5

One digit is
correct but
in the wrong
position

9	3	0

Two digits are
correct but in
wrong positions

8	4	2

One digit is
correct but
in the wrong
position

389.

🔍 63.

→ 809

0	1	2

One digit is correct and in the right position

2	6	4

No digit is correct

3	8	7

One digit is correct but in the wrong position

1	8	3

Two digits are correct but in wrong positions

4	6	9

One digit is correct but in the wrong position

390.

🔍 132.

→ 874

2	9	5

One digit is correct and in the right position

4	9	3

No digit is correct

4	8	3

One digit is correct but in the wrong position

5	7	0

Two digits are correct but in wrong positions

7	1	0

One digit is correct but in the wrong position

391.
🔍 76.
→ 751

8 4 6
One digit is correct and in the right position

6 0 7
No digit is correct

2 3 5
One digit is correct but in the wrong position

2 1 9
Two digits are correct but in wrong positions

9 7 0
One digit is correct but in the wrong position

392.
🔍 17.
→ 468

1 0 5
One digit is correct and in the right position

2 6 5
No digit is correct

6 2 3
One digit is correct but in the wrong position

3 7 4
Two digits are correct but in wrong positions

4 7 9
One digit is correct but in the wrong position

393.

🔍 382.
→ 315

5	0	1

One digit is
correct and
in the right
position

1	8	9

No digit is correct

3	6	7

One digit is
correct but
in the wrong
position

6	7	0

Two digits are
correct but in
wrong positions

8	9	4

One digit is
correct but
in the wrong
position

394.

🔍 293.
→ 629

2	0	8

One digit is
correct and
in the right
position

5	9	8

No digit is correct

5	6	9

One digit is
correct but
in the wrong
position

6	1	0

Two digits are
correct but in
wrong positions

4	1	7

One digit is
correct but
in the wrong
position

395.
🔍 292.
→ 193

7	8	1

One digit is
correct and
in the right
position

9	4	3

One digit is
correct but
in the wrong
position

6	8	0

No digit is correct

2	1	9

Two digits are
correct but in
wrong positions

0	6	5

One digit is
correct but
in the wrong
position

396.
🔍 38.
→ 867

7	5	2

One digit is
correct and
in the right
position

0	8	9

One digit is
correct but
in the wrong
position

5	8	0

No digit is correct

3	2	7

Two digits are
correct but in
wrong positions

4	6	3

One digit is
correct but
in the wrong
position

397.

🔍 269.

→ 175

9 2 8
One digit is correct and in the right position

6 8 4
No digit is correct

3 1 0
One digit is correct but in the wrong position

2 1 5
Two digits are correct but in wrong positions

5 4 6
One digit is correct but in the wrong position

398.

🔍 343.

→ 281

5 0 2
One digit is correct and in the right position

0 4 7
No digit is correct

4 3 7
One digit is correct but in the wrong position

3 9 1
Two digits are correct but in wrong positions

1 9 8
One digit is correct but in the wrong position

399.
🔍 157.
→ 480

8 1 9
One digit is correct and in the right position

7 9 6
Two digits are correct but in wrong positions

1 2 3
No digit is correct

3 4 2
One digit is correct but in the wrong position

7 6 5
One digit is correct but in the wrong position

400.
🔍 99.
→ 367

4 0 3
One digit is correct and in the right position

3 2 7
Two digits are correct but in wrong positions

0 6 9
No digit is correct

6 8 9
One digit is correct but in the wrong position

5 2 7
One digit is correct but in the wrong position

Title:
Solve Hidden Numbers, Volume 1:
Crack 3-Digit Codes Using Hints

Subtitle:
400 Math and Logic Riddles with Tips and Solutions

Edition I

ISBN: 9798301140655

Book Design and Cover: Tobiasz Małysa — Grafista

Contact the Publisher:
www.Grafista.pl   poczta@grafista.pl

Printed in Great Britain
by Amazon